THE
ARMCHAIR
GENERAL

Armchair General Limited

THE ARMCHAIR GENERAL

PROFESSOR JOHN BUCKLEY

CENTURY

1 3 5 7 9 10 8 6 4 2

Century
20 Vauxhall Bridge Road
London SW1V 2SA

Century is part of the Penguin Random House group of companies
whose addresses can be found at global.penguinrandomhouse.com.

Penguin
Random House
UK

First published by Century in 2021

www.penguin.co.uk

A CIP catalogue record for this book is available from the British Library.

ISBN 9781529125702

Typeset in 12/15pt Times LT Std by Jouve (UK), Milton Keynes
Printed and bound in Great Britain by Clays Ltd, Elcograf S.p.A.

The authorised representative in the EEA is Penguin Random House Ireland,
Morrison Chambers, 32 Nassau Street, Dublin D02 YH68

Penguin Random House is committed to a sustainable future for
our business, our readers and our planet. This book is made from
Forest Stewardship Council® certified paper.

MIX
Paper from
responsible sources
FSC
www.fsc.org
FSC® C018179

For Julia, Annabel and Edward

THE
ARMCHAIR
GENERAL

CONTENTS

INTRODUCTION

A few years ago, on a warm sunny day in May, I was standing on the John Frost Bridge in Arnhem in the Netherlands with a group of students from my university's War Studies Department, engaged in a battlefield study tour. We had been tracking the route of the Allies as they fought to seize a crossing over the Rhine in September 1944, part of the infamous Operation Market Garden and graphically depicted in the motion picture *A Bridge Too Far.* Over three days we had covered the route taken by the Allies all the way from Joe's Bridge, situated outside the Belgian town of Neerpelt, through Eindhoven and Nijmegen, finally reaching Arnhem, some sixty miles from the start line of the battle and the scene of the most bitter fighting of the campaign. We had discussed the battles in great detail, examined the key actions, visited the important sites. Now we were drawing to a close on the bridge itself – or the 'prize', as it was called by one commander in 1944 – the primary objective of the battle. One group of students was peering out from the bridge across the Lower Rhine towards the open land to the south and was engaged in heated debate. Why, they wanted to know, had the British not deployed paratroopers there rather than dropping them eight miles away to the west, a huge distance that was ultimately a root cause of the plan's failure? If they had landed here, one student forcefully pointed out, the bridge would have been captured quickly and the whole plan might have proved successful, rather than foundering in bitter acrimony, as it did in 1944. If the British had deployed paratroopers much closer to the bridge in Arnhem, what might the outcome have been? Would success in the battle have followed? Could the war have been won in 1944?

The debate continued for some time without an entirely convincing conclusion being reached, but the discussion highlighted for me once again the role of the crucial pivotal decision in military history, the effect of the commander's choices and the enduring fascination of the impact of alternative outcomes, if those choices and decisions had been different. For some historians this remains little more than an absorbing parlour game and cannot be viewed as real or serious history; after all, the discipline of history concerns itself with interpreting the past through evidence, and there cannot be 'evidence' of things that did not happen. There is enough debate and argument over establishing convincing arguments for why events occurred as they did; dealing with the potential implications of alternatives – or counterfactuals, as they have been labelled – is a further step into the unknown that is fraught with uncertainty, and a step too far for some.

Yet the concept of alternative histories, or 'what ifs', remains enduringly popular in fiction, both in literary and visual media, and a great many of them centre on the outcomes of war or major political events in the 1930s and 1940s. Books such as Len Deighton's *SS-GB* and C. J. Sansom's *Dominion* have explored the much-chewed-over scenario of Britain falling under Nazi hegemony in the Second World War, though it has to be said they are quite light on the process by which this might have occurred. Philip Roth's acclaimed *The Plot Against America* examined an alternative history in which Roosevelt was pipped to the presidency in 1940 by Charles Lindbergh, standing on a platform that was much more sympathetic to the Nazis. The TV series *The Man in the High Castle*, which was set in a USA defeated by the Axis powers in the Second World War and was based on a novel by Philip K. Dick, proved hugely popular and ran for four seasons (2015–19).

Historians have also been drawn to the topic of alternative timelines, with books dedicated to exploring key events throughout history that have shaped our modern world, and the possible impact of different choices being made at pivotal moments. Many have focused on crucial military campaigns and wars – what might have happened if Gabriel Princip had failed to assassinate Archduke Franz Ferdinand in 1914, or if the Duke of Wellington had lost the Battle of Waterloo, for example. Such approaches take us on journeys into alternative possible histories, but they are in essence linear – the reader is completely at the mercy of the author.

Yet for me, although these approaches are interesting and can be useful in understanding why critical decisions may have major repercussions, they do not capture the essence of the intense pressures playing on decision-makers at pivotal moments, particularly during conflicts. I have used gaming methods in teaching for many years to explore how, and on what basis, commanders and leaders make crucial decisions, and the implications – both short- and long-term – of those choices. I have had students conduct research, create grand strategies and devise battlefield plans across a whole series of wars in order to get them to better understand the reasons behind the decision-making process, the pressures at work on commanders, the incomplete data that leaders have to work with and the possibility that the enemy might react in different ways. Opposing teams of students have refought the Agincourt campaign of 1415, key battles in the Thirty Years War, Antietam in the American Civil War, the Battle of Jutland in 1916 and Pacific naval battles of 1942, such as Midway and the Eastern Solomons. All have offered fascinating (and sometimes amusing) insights into command and leadership, and into the psychology and pressures at work, even in teaching rooms insulated from the life-and-death struggles of war. Games and simulations have, of course, been used by the military for centuries as a means of teaching lessons, but also as way of getting officers to think about how and why command decisions are made. Students of history similarly get much from the same approach – frustration and bewilderment sit easily alongside incisive thinking and great perspicacity.

In this book I have brought the notion of 'what if' history together with the essence of decision-making. Each of the scenarios in the chapters ahead is set in the Second World War at a key moment in the fight against the Axis powers. They focus on the defining moments of the war in which crucial and pivotal command and leadership choices were made, both at political and military levels. You will be placed in the position of evaluating evidence, balancing the possible outcomes and making the crucial decision that will lead you either along the historical path or possibly into alternative timelines, with the differing ramifications and potential outcomes of both. Some might yield great success, but others might lead to crushing defeat, with dramatic and calamitous outcomes. As far as possible I have presented the counter-factual alternatives as 'real' historical accounts and have used contemporary pieces of evidence and material to colour the alternative outcome, to add authentic flavour. So in the alternative-history sections Churchill did utter those words, albeit in a slightly different context; and Eisenhower did write that memo, although not quite at that time. I have only tinkered with the history.

You will take on particular roles, often of those people who advised and informed the critical decision-makers – and you will quickly notice that men dominated these decisions even at a political level, a sure marker of the times. You will be directly involved in the strategic and tactical decisions that shaped the course of the war, from the underground Cabinet War Rooms in Whitehall to a general's make-shift command post deep in the Libyan desert, from the Oval Office at the White House to the cockpit of a Douglas Dauntless dive bomber over the Pacific Ocean.

At one moment you might be acting as the personal military assistant to Churchill or Roosevelt, the next as Chief of Staff to Field Marshal Montgomery, then as a chief of security to Stalin, as intelligence analyst for Admiral Nimitz at Pearl Harbor or perhaps even as a scientist whose intimate knowledge of theoretical physics could lead to the development of the atomic bomb. Your influence in these pivotal moments will shape the outcome of the war, and your choices will define the nature of the world that might emerge from the most destructive conflict in history.

I have kept as much as possible to the plausible rather than the fantastical, however. Naturally events have to be tweaked to offer alternatives, and more often than not there were compelling reasons why the historical path was the most obvious and logical; commanders and leaders generally only made foolish decisions that appeared so in hindsight. But I did not inject completely ahistorical elements into the scenarios – no hydrogen bombs in 1940 or F-14s over Pearl Harbor in 1941 (notwithstanding the evidence of the amusing 1980 feature film *The Final Countdown*). Everything that will inform your decision-making as you follow a timeline will appear to be a realistically possible outcome; these will just be paths that did not quite make the cut in history, but could have done so without stretching credibility too far. Hopefully the scenarios will illustrate that the historical paths that were taken were adopted for what appeared to be legitimate reasons at the time, but that the alternatives often made some sense too.

Yet I am convinced that however much I make the argument (as I did to my students during our trip to the Netherlands) that the decision in 1944 to drop paratroopers eight miles from the bridge at Arnhem *was* logical, others analysing the battle may think it a decision that would have been better consigned to the list of implausible alternatives.

RULES OF ENGAGEMENT: NOTES ON READING THIS BOOK

1. The eight chapters in this book span scenarios that shaped the Second World War. Each focuses on a series of key decisions and choices that profoundly altered the course of the war and might have resulted in significant changes to the outcome of the conflict. Scenarios have been selected that best highlight chains of potentially decisive verdicts.

2. Each scenario is divided into sub-scenarios, and within each scenario you will face a series of crucial decisions that will require you to make an informed but potentially critical intervention, in the form of a multiple-choice question requiring your action. You must decide which course of action to take, then turn to the appropriate page.

3. Each chapter starts with a different historical premise taken from the war itself, and each will potentially lead you into uncharted historical waters.

4. Each choice you make will lead either to a further decision or to an outcome. Further choices could be based on the historical path – some on the possibilities opened up by your decision-making. One outcome will be the historical path, while others will outline the alternative based on your choice.

5. To help inform you in your decision-making there is a series of primary source documents, including maps, meetings minutes and data that decision-makers at the time would have been provided with. Read and analyse them carefully, but do not always accept what they say.

6. Good luck in your reading and in your mission ahead. As Winston Churchill said: 'Fear is a reaction; courage is a decision.'

PROTAGONISTS

1 BRITAIN'S DARKEST HOUR

You are:

David Margesson, MP, the Conservative Party's Chief Whip and Chamberlain's enforcer of discipline (p. 14).

General Hastings 'Pug' Ismay, Churchill's trusted military advisor and his liaison officer with the military Chiefs of Staff (p. 24).

Clement Attlee, MP, leader of the Labour Party (p. 34).

Samuel Hoare, Foreign Secretary (p. 44).

Anthony Eden, Foreign Secretary (p. 50).

Alexander Cadogan, the Permanent Under-Secretary at the Foreign Office (p. 56).

Archie Sinclair, Secretary of State for Air (p. 62).

2 THE MEDITERRANEAN GAMBLE

You are:

Brigadier Eric 'Chink' Dorman-Smith, an Irish officer in the British Army, working for General Archibald Wavell, the Commander-in-Chief of the Middle Eastern theatre (p. 72).

Jacques Tarbé de Saint-Hardouin, the political assistant to General Maxime Weygand (p. 80).

General Alan Brooke, the fearsome Chief of the Imperial General Staff (CIGS), in effect the head of the entire British Army (p. 88).

İsmet İnönü, the Turkish President (p. 96).

Admiral Andrew Cunningham, the highly regarded Commander-in-Chief of the British Mediterranean Fleet (p. 100).

Brigadier Francis 'Freddie' De Guingand, Montgomery's Chief of Staff (p. 106).

Lieutenant Colonel Charles Richardson, the chief planning officer at the Eighth Army (p. 112).

3 STALIN'S WAR

You are:

Vyacheslav Molotov, the Foreign Minister of the USSR (p. 122).

Nikolai Vlasik, Stalin's chief of personal security (p. 130).

General Georgy Zhukov, Chief of the General Staff (p. 136).

Anastas Mikoyan, the State Defence Committee member responsible for supplying the armed forces with food and war materiel (p. 142).

Boris Rybkin / Yartsev, a secret agent of the Soviet Union's NKVD (p. 148).

Colonel General Pavel Artemyev, the Chief of the Moscow Military District (p. 156).

Christian Günther, Sweden's Foreign Minister (p. 162).

4 MIDWAY – DECISION IN THE PACIFIC

You are:

Captain Lynde D. McCormick, chief planning officer to the US Navy's Pacific Fleet Commander-in-Chief, Admiral Chester Nimitz (p. 174).

Captain Miles Browning, Spruance's Chief of Staff and Air Tactical Officer on the USS *Enterprise* (pp. 184, 200).

Captain Elliott Buckmaster, the commander of the aircraft carrier USS *Yorktown* (p. 192).

Lieutenant Commander Wade McClusky, the leader of the USS *Enterprise*'s air group (pp. 204, 212, 218).

5 BOMBER OFFENSIVE

You are:

Lord Beaverbrook (Max Aitken), a Canadian press baron (p. 226).

Arthur Street, the Permanent Secretary at the Air Ministry (p. 234).

Admiral of the Fleet Dudley Pound, the First Sea Lord and Chief of Naval Staff. (p. 242).

Air Vice Marshal Norman Bottomley, the Deputy Chief of Air Staff (DCAS) (pp. 248, 262).

Air Commodore Sidney Bufton, the Director of Bomber Operations (p. 256).

Air Vice Marshal Norman Bottomley, the newly appointed Air Officer, Commander-in-Chief, Bomber Command (p. 262; see also p. 248).

Air Vice Marshal Robert Saundby, Arthur Harris' deputy Commander-in-Chief of Bomber Command (p. 267).

6 CASABLANCA

You are:

Field Marshal John (Jack) Dill, Chief of the British Joint Staff Mission based in Washington and Churchill's senior representative on the CCS (pp. 276, 322).

General Dwight D. Eisenhower, Supreme Commander Allied Expeditionary Force for the North African Theatre (p. 286).

General Walter Bedell Smith, Marshall's Chief of Staff (p. 294).

Dick Mallaby, a member of Churchill's famous Special Operations Executive (SOE) (p. 302).

General Henry 'Jumbo' Maitland Wilson, Commander-in-Chief, Middle East (p. 310).

General Harold Alexander, commander of the greatest amphibious invasion in history, Operation Roundup (p. 316).

7 RACE TO THE RHINE

You are:

Air Marshal Arthur Tedder, Eisenhower's Deputy Supreme Commander (p. 330).

General Lewis Brereton, the recently appointed commander of FAAA (p. 340).

Lieutenant General Frederick 'Boy' Browning, the dashing leader of I British Airborne Corps (p. 346).

Lieutenant General Brian Horrocks, the commander of XXX Corps (p. 354).

Lieutenant General Miles Dempsey, the quiet and unassuming commander of Second British Army (p. 360).

Brigadier Edgar 'Bill' Williams, 21st Army Group's chief of intelligence (p. 366).

Brigadier General Floyd Parks, Brereton's Chief of Staff (p. 372).

8 THE BOMB

You are:

Margrethe Bohr (née Nørland), Niels Bohr's wife and key confidant (pp. 383, 398).

Vannevar Bush, the dynamic and determined head of the US Office of Scientific Research and Development (OSRD) (p. 388).

Henry Stimson, the US Secretary of War (p. 405).

General George Marshall, the head of the US Army (p. 412).

Vyacheslav Molotov, the USSR's Foreign Minister (p. 422).

Colonel Paul Tibbets of the 509th Composite Group of the United States Army Air Force (p. 426).

1

BRITAIN'S DARKEST HOUR:
SUMMER 1940

SECTION 1
THE PARTY'S DILEMMA

You have sat too long here for any good you have
been doing. Depart, I say, and let us have done with
you. In the name of God, go!

LEO AMERY, MP

7 May 1940

There is hardly an empty seat as you listen to the MP Leo Amery's
fiery words, directed at the Prime Minister Neville Chamberlain, a
man whom hitherto Amery had supported and considered to be a
friend. You are ***David Margesson***, MP, the Conservative Party's Chief
Whip and Chamberlain's enforcer of discipline. As such, you are a
key power broker in Parliament, feared by many in the party as a
tough and ruthless disciplinarian, a man of resolve and grit. But as
you sit on the green leather benches of the House of Commons, listening intently to the unfolding debate about the government's handling
of the war effort, your concern begins to grow. As Chief Whip, it is
essential that you have a feel for the mood of the House, and your
instincts are telling you that all is not well. Britain's situation in the
war against Hitler's Germany is unravelling and so too, it seems, is the
integrity of Chamberlain's government.

Over the next few days the House of Commons, this grand political theatre, will be the stage for one of the most momentous wartime
verdicts ever made in British politics. Amery's speech, alongside open
criticism from many other key military and political figures, is setting

in motion a chain of events that will see Chamberlain unseated in a matter of days and his government replaced by a national coalition. Sporting old-fashioned wing-collars and decidedly antiquated garb, Chamberlain appears increasingly out of touch. His speech during the debate is deeply unimpressive and gives the sense of a 'shattered man'. But the dilemma facing you – one of the key players in Parliament – is: Can Chamberlain's premiership survive and, if not, who in the party could possibly replace him?

1. David Margesson, MP

THE NORWAY CAMPAIGN

Leo Amery's words continue to reverberate around you. After an ill-tempered day of debate, the House is entering the tumultuous closing stages of a post-mortem analysis on the disastrous Norway campaign, which has seen thousands of Allied soldiers killed and numerous others rendered powerless in the face of Hitler's blitzkrieg. The Allied forces sent to intervene in Norway have proved singularly unable to influence events, and the British have lost more than 4,000 soldiers, sailors and airmen, to say nothing of more than a hundred invaluable aircraft, seven destroyers, two cruisers and an aircraft carrier. In truth, Winston Churchill, who was brought back into government as First Lord of the Admiralty in the autumn of 1939 by Chamberlain himself, is in some ways more responsible for the fiasco. It had been his idea to intervene in Norway as part of a grander vision of isolating Scandinavia from Germany. Yet blame does not fix upon him, and it is Chamberlain who is in the cross-hairs. Roger Keyes, a Conservative MP and a serving admiral (wearing his uniform in the Commons for the first time to emphasise the point), delivers a poignant speech

I came to the House of Commons to-day in uniform for the first time because I wish to speak for some officers and men of the fighting, sea-going Navy who are very unhappy.

The Prime Minister told the House last week the objectives which the Government had in view in the Norwegian campaign, and he said: 'It was obvious that these objectives could be most speedily attained if it were possible to capture Trondheim, and, in spite of the hazardous nature of the operation, with the Germans in possession of the place and in occupation of the only really efficient aerodrome in South West Norway, at Stavanger, we resolved to make the effort.'

Immediately the Norwegian campaign opened I went to the Admiralty to try and suggest action, based on my considerable experience of combined operations in amphibious warfare in the Dardanelles and on the Belgian coast. I was foolish enough to think that my suggestions might be welcomed, but I was told it was astonishing that I should think that all these suggestions had not been examined by people who knew exactly what resources were available, and what the dangers would be.

When I realised how badly things were going later, and saw another Gallipoli looming ahead, I never ceased importuning the Admiralty and the War Cabinet to let me take all responsibility and organise and lead the attack.

Once again there is a deadlock on the Western Front. If only we had used our sea-power vigorously and courageously, the German army in Norway would by now be in a very dangerous position, and would eventually have been decisively defeated.

Source 1: Admiral Roger Keyes' speech, Hansard, 7 May 1940 (extract) outlining the haplessness of the forces committed to battle without adequate support and equipment. A loyal party man, his words hurt the government hard, however inelegantly they are delivered (see Source 1). As one witness put it, 'The sincerity that lay behind his words gave them life.'

8 May 1940

It is Clement Attlee, leader of the Labour Party, who forces the issue. Dismayed by yesterday's debate, he decides to press for a no-confidence vote in the Prime Minister, not in the expectation of winning, but in the hope of damaging Chamberlain and forcing him to reshape his government. To some in the House of Commons, that Chamberlain has remained as Prime Minister after the outbreak of war in September 1939 has been a farce; he was, after all, intimately involved in the policy of appeasement – a policy that hinged on pre-venting war with Germany by doing deals with Hitler. The most obvious and egregious example was, of course, the disastrous Munich Agreement of 1938, which sold out Czechoslovakia to avoid conflict with the Nazis. Even then, when Germany invaded Poland a year later, Chamberlain still clung to the idea of a negotiated settlement. It took a near Cabinet rebellion to persuade him to deliver the ultimatum to Hitler that eventually took Britain to war. Attlee is doubtful that he can unseat Chamberlain, but hopes to rid the government of some dead wood and join a new, fully fledged national coalition.

As the confidence vote looms, you work tirelessly to bolster sup-port for the government, resorting to threats when guile does not produce success. In the event, the government carries the day in the vote, but sheds a crippling amount of support; its theoretical majority of 213 in the House is cut to just eighty-one. Many Tories abstain, but some forty-one actually vote against their own government. When you find out that your fellow Conservative John Profumo has voted with the opposition you let him have both barrels, declaring him an 'utterly contemptible little shit!' Despite your best efforts, the Prime Minister's position has been severely compromised and although

Chamberlain is unlikely to resign just yet, the matter of who might replace him – if, and when, he does – is now urgent. It is your job to take soundings and estimate the level of support for his possible replacements.

THE CANDIDATES

Winston Churchill

To any observer in Westminster, Churchill cuts a divisive figure and has quite limited support. It is true that he campaigned tirelessly against appeasement in the 1930s and that he has a big national media presence and profile; his speeches, and his apparently unflinching desire for war against the Nazis, stand him in good stead. But he also carries considerable political baggage – to those on the left, he will forever be tarnished by the Sidney Street Siege of 1911 when, as Home Secretary, he had ordered soldiers onto the streets of London to engage in gun battles with some Latvian revolutionaries, an incident that was seen as heavy-handed, to say the least. He is also perceived as 'damaged' by Labour for his strenuous efforts to crush the General Strike of 1926, and for his openly positive advocacy of the Empire. Labour MPs had resolved never to work with 'such a villainous character as Churchill'. To those on the right of politics, he is the man who had abandoned the Conservatives to join the Liberals and had then slunk back when the Liberal Party imploded in the 1920s. In the 1930s, during his wilderness years when excluded from high office, he had then spent much of his time hammering away at his own party's government on a range of high-profile issues. To some people, Churchill is the man responsible for the Gallipoli fiasco, the failed campaign that he devised in 1915 to drive Ottoman Turkey out of the First World War, but which instead resulted in the deaths of nearly 60,000 Allied soldiers. To others, Churchill is simply a hot-headed, impetuous maverick.

For these reasons, he does not appear to be a popular choice to replace Chamberlain, particularly within his own party. Leading Tory figures such as Stanley Baldwin, Max Beaverbrook and Sam Hoare are sure that the Conservative Party would not tolerate Winston as leader. Other names are bandied about: Anthony Eden or even the now seventy-seven-year-old David Lloyd George, the victor of the First World War, but none are convincing – none, that is, other than the Foreign Secretary, Edward Wood, or Lord Halifax as he has been since 1934.

Lord Halifax

Halifax, known by some as the 'Holy Fox' for his political acumen and guile, is a highly regarded figure in British politics, having held a range of senior positions, including Secretary of State for War and Viceroy of India. Since early 1938 he has been Foreign Secretary, and although he is implicated in Chamberlain's appeasement policy, he was decidedly cool to the Munich Agreement and was a key figure in pressing Chamberlain to stand firm over Poland. Urbane, erudite, and with an aristocratic bearing described as 'a lofty moral principle', Halifax is also slim and very tall (6' 5"), and thus towers over the portly 5' 7" Churchill. Even a deformed arm and an inability to pronounce his 'r's has not held him back in life (see Picture 2).

2. Churchill and Halifax

Some regard Halifax as 'a tired man' and naturally lazy, but in the aftermath of the Norwegian debate, if change is required – for Chamberlain and many other Tories – then Halifax is the natural successor and the man to take over. Even the opposition parties appear to be willing to work with him; he has, after all, worked under a Labour government when Viceroy of India without difficulty. If a government of national unity is now required, Halifax appears the man most likely to gain the trust of the Labour and Liberal Parties. Churchill has the wider popular support outside Parliament, but in the corridors of power it is Halifax who holds the best hand; and after the Norway debate you understand that Chamberlain is pressing Halifax to be ready to step up, if required.

But Halifax is hesitant for two reasons. First, he sits in the House of Lords and would not be able to debate and lead in the Commons, even as Prime Minister. The problem could be overcome if emergency legislation were quickly passed to allow him to return to the Commons, but there might be something of a hiatus while it is enacted. In truth, this is not a real problem and could be overcome, but it offers Halifax an 'official' reason not to take the post of Prime Minister. But the real issue troubling Halifax, you have heard, is that he is fearful of taking the post – 'sick to pit of the stomach' even. He claimed reticence at becoming Foreign Secretary in 1938 and has often feigned unease at taking senior and responsible posts, wanting to be assured of overwhelming support. In classic aristocratic stereotype, he does not want to be seen desiring high office, but to be someone 'dragged there by friends for his country's good', when in truth he desires a quiet country life. Is he therefore up to the task ahead?

9 May 1940

Chamberlain invites Churchill, Halifax and you, as the Chief Whip, to meet him at Downing Street in the late afternoon, less than a day after the damaging vote in the Commons, to decide a way forward. You have told Halifax over the previous twenty-four hours that he has the backing of the Conservative Party hierarchy, the opposition leaders,

Chamberlain himself and even the King – like many others, the King naturally assumes that Halifax, a man whom he and his wife much prefer, will take up the post of Prime Minister. Naturally Churchill's attitude is an issue: will he support Halifax as Prime Minister? Or does he see himself in the role – something that, despite reservations, you think might also be a good solution. Perhaps most of all, what matters now is Halifax's own determination to take up the reins of power.

At first Chamberlain tries to gain the support of Halifax and Churchill for him staying on as Prime Minister, but you have to intervene: Clement Attlee has already told you that he will almost certainly not bring the Labour Party into a new national government headed by

3. Churchill and Chamberlain

Chamberlain – so your current leader will inevitably have to step aside. Visibly deflated, Chamberlain turns to the others for their views. Churchill is uncharacteristically silent; Halifax hesitant. If and when Chamberlain finally resigns, which of the two leading candidates will you support?

AIDE-MEMOIRE

* Halifax is a hugely experienced figure who has been in the Cabinet for much of the 1930s.

* Churchill has high-level political experience dating back to before the First World War.

* Halifax would convey calm and assured authority in this time of crisis and uncertainty.

* Churchill would bring dynamism and energy to the leadership of the government.

* Halifax has the backing of the senior party leadership and the tacit support of the opposition parties, who need to be brought on board.

* Churchill has popular support in the country, essential to rallying support.

* Halifax is hesitant and appears to lack desire for the job. Surely self-belief and drive will be essential in the months to come?

* Churchill can be erratic and reckless (consider Gallipoli and Norway). Britain needs clear, analytical leadership rather than impulsive and costly disasters.

THE DECISION

Should you support Halifax or Churchill?

▶ **Yes** to Churchill becoming Prime Minister, go to **Section 2** (p. 24).

OR

▶ **Yes** to Halifax becoming Prime Minister, go to **Section 3** (p. 34).

SECTION 2
THE HINGE OF FATE – CHURCHILL TAKES CHARGE

25 May 1940

The War Cabinet is once again assembling in the gloomy underground Cabinet War Rooms below Whitehall, against a backdrop of depressing news about Britain's deteriorating war effort. Across the English Channel the Allied armies are in a perilous position and the country faces possible total disaster. It is also becoming increasingly clear that, with the nation on the brink, some elements are starting to question whether new Prime Minister Winston Churchill's bullish and uncompromising approach to the conduct of the war is out of step with reality. As the leaders of Britain's government settle into their seats to thrash out a way forward, the mood is sombre; the next few days could well be a decisive turning point in the whole war.

You are *General Hastings 'Pug' Ismay*, Churchill's trusted military advisor and his liaison officer with the military Chiefs of

4. General Hastings 'Pug' Ismay

Staff, the heads of the Army, the Royal Air Force and the Royal Navy. Since Churchill became Prime Minister you have been sitting in on many War Cabinet Defence Committee meetings and have clear insights into the major decision-making of the British government at the highest level. The Prime Minister relies on your advice and comments, and you sense that over the next few tumultuous days he will be depending on you even more.

It had all seemed so different a little over two weeks earlier when Churchill had emerged from the meeting at Downing Street on 9 May as the man chosen to become the next Prime Minister, if and when Chamberlain resigned. You have heard that Chamberlain tried to stay on, but that the Labour Party made it abundantly clear that a change was needed. Remarkably Lord Halifax, the most likely choice to become Prime Minister, had hesitated. Perhaps he had been worried about how much he would actually be able to lead the country with the high-profile Churchill running the war as Minister of Defence under him; or perhaps he was simply nagged by the view that Winston might well indeed be the right man for the job. Apparently Churchill had remained silent as Halifax pondered, prompting his rival to worry that he was unwilling to support him; this perhaps caused Halifax to back down and throw his support behind Churchill. Despite their reservations, Attlee and the Labour Party soon agreed to Churchill taking over, with Halifax staying on as Foreign Secretary.

In the evening of 10 May Churchill met a hesitant George VI to accept the King's request to become the new Prime Minister. Considering the enormity of the task ahead of him, Churchill ruminated, 'I hope it is not too late. I am very much afraid it is.'

By the time the appointment was formalised, German forces had launched an all-out offensive against the Low Countries and France, plunging Western Europe into crisis. It was an inauspicious moment, to say the least. On 13 May Churchill offered the nation his 'blood, sweat and tears' speech in pursuit of victory, though many Tories in the House of Commons were less than impressed by such rhetoric. As Churchill spoke, the German blitzkrieg was striking Allied forces on the continent a decisive blow and inflicting crippling defeats upon them; the Netherlands capitulated on 15 May and Belgium appeared to be next. The situation had become grave in the extreme when German tanks, troops and aircraft had forced a crossing over the River Meuse at Sedan a few days before, on 12 and 13 May, broke through Allied lines and thrust towards the English Channel, threatening to cut the Allied armies in two. Pandemonium broke out in Paris and London.

On 15 May the French Prime Minister, Paul Reynaud, had telephoned Churchill to announce, 'We have been defeated, we are beaten.' Churchill flew to Paris the next day and witnessed for himself the despair and panic amongst the French leadership and was unable to inspire either hope or a new plan. Sporadic counter-attacks by Allied forces had done little to halt the German drive to the English Channel, which reached the coast on 21 May, splitting the Allied armies apart; the British and some French forces went to the north, the bulk of the French army to the south. British forces were being pushed further and further back towards the coast around Calais and Dunkirk and defeat appeared inevitable.

Churchill was embattled at home, too. Defeatism and woe swept through the corridors of power in London. 'This is the end of the British Empire,' claimed General Ironside, Chief of the Imperial General Staff. Churchill's appointment as Prime Minister had not gone down well with the majority of Tories in Westminster in the first place; now they slumped in despair. What could this 'unsound' and 'unstable' maverick achieve?

In this climate Churchill is also having to confront a growing feeling from some in his government, and indeed his War Cabinet, that Britain might be best served by getting out of the war before total calamity sweeps everything away. There even exists in some quarters the belief that Britain is fighting the wrong war anyway and that Stalin's Soviet Union is the real enemy, representing a far greater threat than Germany. Hitler is, it is claimed by some people, surely more interested in striking east than west?

Perhaps now is therefore the moment to seek a way out of the war while Britain still has some credibility? The British Expeditionary Force on the continent is just about intact, the French are still fighting (however ineptly some in London believe) and the Germans, though in the ascendancy, are still unsure of outright victory. What might Hitler accept as the price for France and Britain withdrawing from the war? Direct negotiations with the Germans are not possible, but the Italian leader Benito Mussolini offers a tentative diplomatic route to Berlin. The Italians are still neutral, and some in the British government believe that Mussolini is wary of Hitler seizing too much power in Europe and might therefore be willing to broker a deal. At the very least the Italians might be prevailed upon to stay out of the war.

You, as General Ismay, look on as Churchill's War Cabinet begins to debate the situation. Churchill has established a small War Cabinet of five senior figures, headed by himself as Minister of Defence as well as fulfilling the role of Prime Minister (thus making him the nearest thing to a military dictator that the country has witnessed since Oliver Cromwell). Neville Chamberlain now acts as Lord Chancellor and still leads the Conservative Party, thus making it vital that Churchill retains his support. Also sitting around the table are Clement Attlee, the Labour leader, and his deputy, Arthur Greenwood. Attlee is infamously quiet and a rather colourless figure – 'a timid mouse' as one put it – but he has been pivotal in Labour taking a pro-rearmament / anti-appeasement stance against Hitler since the late 1930s. Greenwood, a sixty-year-old working-class Yorkshireman, is more outgoing and genial, but there are rumours that he is a heavy drinker and not always as resolute as he should be. As part of the new national coalition government, their support and influence are vital.

Churchill has also retained Lord Halifax as Foreign Secretary, partly because he respects his abilities and experience (though not always his judgement) and partly because he knows that Halifax retains considerable support in the Conservative Party. These five men effectively make the big decisions now, supported and advised, as and when required, by civil servants, military personnel, other invited ministers and, in Churchill's case, you. Oddly, in the War Cabinet Churchill probably has more support from Labour's Attlee and Greenwood to bolster his position than from Halifax and Chamberlain, who you suspect are still smarting about the manner in which Churchill became Prime Minister.

As the War Cabinet meeting progresses, Halifax, in his capacity as

Foreign Secretary, announces that he had been in contact with the Italian ambassador, Giuseppe Bastianini, to begin exploring the price not only of keeping Italy out of the war, but also of possibly widening the talks into a general European settlement. Churchill appears content at this stage to see if the contact might at least buy Britain some time with Italy, but is clear that the meeting 'must not . . . be accompanied by any publicity, since that would amount to a confession of weakness'. Halifax comments that he will meet Bastianini that afternoon.

Sunday 26 May 1940

The news from the continent is gloomier than ever. Rumours are rife that Belgium is about to surrender, while Paul Reynaud, the French Prime Minister, is flying over to London – quite possibly, you suspect, to seek a way of getting France out of the war. The mood in the War Cabinet as it meets again is downbeat. Churchill tries to shore up the position of carrying on the war in spite of everything; he tables an optimistic paper prepared by the military Chiefs of Staff, which argues that Britain could survive and carry on even if France threw in the towel (see Document 1 on pp. 29–30).

Halifax, however, ups the ante and brings up his meeting with the Italians. He presses the War Cabinet to agree to allow him to take the matter further. What terms would Mussolini accept to stay out of the war, or could he even open discussions with the Germans for a wider settlement? Halifax argues that it is essential to find out these matters as quickly as possible. Churchill frowns and counters that any terms on offer from Germany and Italy could never be acceptable, and that to open up any dialogue would send out all the wrong signals. He is fearful that negotiations of any kind led by Halifax might be the thin end of the wedge, leading to a humiliating Munich-style 'deal'. Churchill confides to you that any deal could only result in Britain being reduced to vassal status.

But Churchill's strategy and long-term vision of how Britain might prevail in a war where France has collapsed are less than convincing; they appear to boil down to mere survival, with the lingering hope

330

TO BE KEPT UNDER LOCK AND KEY.

It is requested that special care may be taken to
ensure the secrecy of this document.

SECRET.

W.P.(40)169.

COPY NO. 17

(Also Paper No.
C.O.S.(40) 397).

26TH MAY, 1940.   WAR CABINET.

BRITISH STRATEGY IN THE NEAR FUTURE.

Report by the Chiefs of Staff Committee.

We have reviewed our Report on "British Strategy in
a Certain Eventuality" (Paper No. W.P. (40) 168) in the
light of the following Terms of Reference remitted to us
by the Prime Minister.

"In the event of France being unable to continue in
the war and becoming neutral with the Germans holding
their present position and the Belgian army being forced to
capitulate after assisting the British Expeditionary Force
to reach the coast; in the event of terms being offered
to Britain which would place her entirely at the mercy
of Germany through disarmament, cession of naval bases in
the Orkneys etc; what are the prospects of our continuing
the war alone against Germany and probably Italy. Can
the Navy and the Air Force hold out reasonable hopes of
preventing serious invasion, and could the forces gathered
in this Island cope with raids from the air involving
detachments not greater than 10,000 men; it being observed
that a prolongation of British resistance might be very
dangerous for Germany engaged in holding down the greater
part of Europe."

2.   Our conclusions are contained in the following
paragraphs.

3.   While our air force is in being, our Navy and

air force together should be able to prevent Germany

carrying out a serious seaborne invasion of this

country.

6.   The crux of the matter is air superiority. Once
Germany had attained this, she might attempt to sub-
jugate this country by air attack alone.

9.   Whether the attacks succeed in eliminating the
aircraft industry depends not only on the material
damage by bombs but on the moral effect on the
workpeople and their determination to carry on in the
face of wholesale havoc and destruction.
10.   If therefore the enemy presses home night
attacks on our aircraft industry, he is likely to
achieve such material and moral damage within the
industrial area concerned as to bring all work to a
standstill.

11.   It must be remembered that numerically the Germans
have a superiority of four to one. Moreover, the German
aircraft factories are well dispersed and relatively
inaccessible.
12.   On the other hand, so long as we have a counter-
offensive bomber force, we can carry out similar attacks
on German industrial centres and by moral and material
effect bring a proportion of them to a stand-still.
13.   To sum up, our conclusion is that prima facie Germany
has most of the cards; but the real test is whether the
morale of our fighting personnel and civil population will
counter balance the numerical and material advantages
which Germany enjoys. We believe it will.

J.G. DILL.   (Signed)   C.L.N. NEWALL.
T.S.V. PHILLIPS.   DUDLEY POUND.
R.E.C. PEIRSE.   EDMUND IRONSIDE.

that the Americans will intervene at some point. In contrast, Halifax wants to pursue the idea of a smaller prize – one that saves something for Britain, but is likely to give away much. The meeting appears deadlocked.

Yet you know that Churchill's position is finely balanced. If he blocks Halifax completely, it might cause him to resign, and that might provoke Chamberlain to walk away, too. With them would go the support of the Conservative Party, and Churchill's position might well become untenable; his government could fall, even if he retains Attlee's and Greenwood's support. You notice that Chamberlain has not yet committed himself either way. The atmosphere is electric, but Churchill pulls off a masterstroke and buys himself a little time. He agrees to Halifax preparing a memorandum to outline the possible ways in which the Italians might be approached. But it is only knocking the decision into the long grass, as it is almost certain that Halifax will want to discuss it openly in the following day's War Cabinet meetings. As the meeting breaks up, you begin to discuss the dilemma with the Prime Minister. What should you advise?

AIDE-MEMOIRE

* The battle on the continent is going very badly. It is quite possible that the entire British Army in France might be lost, unless you act now.

* Consider the Chiefs of Staff report, which suggests that Britain could survive even if France surrenders (see Document 1).

* Failure to get out of the war now might cause the collapse of the British state. Isn't it better to hold on to something, even if the price is some overseas assets and territories?

* Sending out any signal to Berlin that Britain is willing to negotiate will probably make an outcome such as a Munich-style deal much more likely. Appeasement is surely dead?

* It may be better to cut one's losses and negotiate while there is still time. Maybe Halifax is correct that simply exploring the option with Mussolini is the least-bad option?

* Churchill is surely correct that any deal with Germany is a greater risk to Britain in the long term; Hitler has broken all his promises and will just choose to attack again when it suits him.

* If Churchill blocks Halifax too readily, he might resign and possibly bring down the government.

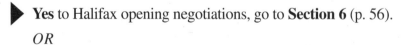

THE DECISION

Should you advise Churchill to foil Halifax and risk him resigning, or should you suggest that he allows Halifax to open negotiations with the enemy?

▶ **Yes** to Halifax opening negotiations, go to **Section 6** (p. 56).

OR

▶ **No** to Halifax opening talks, go to **Section 7** (p. 62).

SECTION 3
THE HOLY FOX'S WAR

25 May 1940

You are **Clement Attlee**, MP, leader of the Labour Party and a key figure in Lord Halifax's War Cabinet. You were brought into government a little over two weeks ago following the resignation of Neville Chamberlain, which in many ways you forced. Since then, as part of Lord Halifax's national coalition government, you have been involved at the highest level in some of the most momentous decisions made in British history. But now, as the War Cabinet assembles in the gloomy underground Cabinet War Rooms deep below Whitehall, a deep sense of despair has taken hold. The war is going very badly, and calamity and defeat appear to be looming fast. Halifax appears weary and despondent as you all start to discuss the deteriorating situation on the continent.

Since 10 May, when Halifax was sworn in as Prime Minister, you have increasingly pondered whether he was the correct choice after all. The Labour Party had been instrumental in removing the hapless and politically damaged

5. Greenwood and Attlee

Chamberlain – the price of your willingness to serve in a new national coalition government – and had accepted that Lord Halifax take up the reins of power. But this had not been without upheaval, because Arthur Greenwood, your deputy, refused to serve in a Halifax government and had stood down, to be replaced by Herbert Morrison, who found the new Prime Minister more palatable. But it had still not been plain sailing; Halifax had insisted upon retaining a number of what you regard as undesirables in the War Cabinet, men such as Sam Hoare and John Simon, both deeply tainted by appeasement and the failed foreign policies of the 1930s.

Halifax had been immediately confronted by a rapidly deteriorating situation on the continent. The German blitzkrieg descended upon the Low Countries on 10 May just as he had been firming up his Cabinet, and within days the Netherlands had surrendered, Belgium was teetering on the brink and France was thrown into ferment. German forces thrust across the River Meuse at Sedan on 14 May, shattering French forces in the area, before breaking out towards the English Channel. By 21 May the Allied position was in tatters. The French government bemoaned impending defeat, while Britain's Chief of the Imperial General Staff, General Edmund Ironside, recently declared that 'At the moment it looks like the greatest military disaster in history' (see Map 1 on p. 36).

Winston Churchill had been installed as Defence Minister under Halifax, with the brief to coordinate and lead the military war effort. Churchill had probably wanted the top job himself, but once Halifax had seized his moment, Winston had to accept the next-best post, for the time being anyway. You had grave reservations about Churchill, but having seen him in action over the last two weeks you have been impressed by his energy and drive. He has been everywhere, flying to and fro across the English Channel to confer with his generals, even though some found his interference exasperating. General Pownall commented, 'Can nobody prevent his trying to conduct operations himself as a super commander-in-chief?' Churchill had also flown to Paris to meet General Maxime Weygand, the new head of the French army. He reported back that he had found despondency and defeatism everywhere, however, and there was even talk of General Philippe Pétain – the eighty-four-year-old First World War hero and known

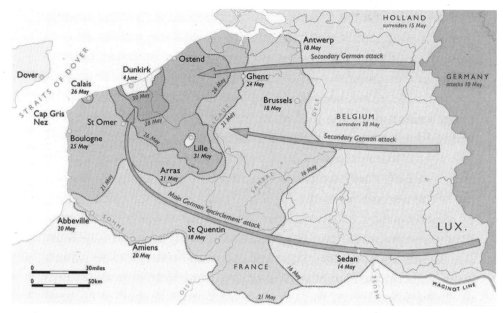

Map 1: German advance, May 1940

opponent of war with Germany, and maybe even a Fascist sympathiser –
taking power.

Back in London, Halifax was also confronting growing despair
and defeatism among the elite of British society, although the extent
of the calamity unfolding on the continent had thus far been concealed
from the wider British public. Much hope had been pinned on a
French-led counter-offensive against the German forces that had battled their way through Allied lines all the way to the English Channel.
Churchill had talked up the chances of success, but the offensive never
amounted to anything substantial, and morale slumped.

France looks doomed, and Halifax has begun to seek ways out of
the unfolding disaster for Britain. A military solution might be increasingly unlikely, but diplomacy potentially offers a route to some form
of short-term security. On taking power, Halifax appointed Samuel
Hoare as his replacement at the Foreign Office, much to the irritation
of you and your colleagues, who wanted Hoare removed from office
completely. Hoare is a known fierce opponent of Churchill and though
he had been at the Air Ministry on 10 May, when Churchill became
Defence Minister, Churchill insisted on Hoare being moved on. Now

at the Foreign Office, Hoare, at the behest of Halifax, has recently opened up communications with Mussolini's Foreign Minister (and son-in-law) Count Ciano, via the Italian ambassador in London, Giuseppe Bastianini.

So far Hoare has been working to prevent Italy joining the war on Hitler's side, but as you look on in stunned silence, it appears that he and Halifax have been up to quite a lot more. The War Cabinet listens intently as Hoare presents a draft document, quickly referred to as the 'Hoare–Ciano Plan', which outlines the Italian demands for staying out of the war and persuading Hitler to call a halt to the battles raging on the continent. Clearly Mussolini has been in touch with Hitler and the outline proposals have the whiff of a deal rubber-stamped in Berlin. Britain would be allowed to bring its armed forces that are facing destruction on the continent back to England, but France would have to accept partition, with the coastal zones along the English Channel and the Atlantic coastline being directly occupied by the Germans. Britain would also have to surrender Malta to Italy and demilitarise Gibraltar, while the German colonies seized at the end of the Great War would have to be returned by the Allies to Hitler. This would secure a ten-year peace treaty (see Maps 2 and 3 on p. 38).

The War Cabinet is stunned by the implications of the Hoare–Ciano Plan; it concedes defeat, but at least salvages something from the wreckage, most obviously the British Army on the continent. Neither Hoare nor Halifax is immediately willing to commit openly to the plan, and both Halifax and Chamberlain outline the severe risks in accepting such a deal. But it would buy time for Britain and prevent total defeat, which now seems imminent; France too would rescue something. The atmosphere in the meeting darkens as Churchill rages at Hoare for his supine 'surrender' terms. He tables his own document, a new Chiefs of Staff report (see Document 2), which argues that, even if France collapses, Britain would still be able to carry on the war.

Churchill scoffs at the idea of a ten-year treaty with the Nazis; how could anyone possibly believe that Hitler would keep to any terms of a deal? It would surely be just like the one the Germans signed with Poland a few years earlier and ripped up at the first

Maps 2 and 3: Territories to be taken by the Axis powers

Map 2

Map 3

330

THIS DOCUMENT IS THE PROPERTY OF HIS BRITANNIC MAJESTY'S GOVERNMENT

TO BE KEPT UNDER LOCK AND KEY.

It is requested that special care may be taken to
ensure the secrecy of this document.

SECRET.

W.P.(40)169.

COPY NO. 17

(Also Paper No.
C.O.S.(40) 397).

WAR CABINET.

26TH MAY, 1940.

BRITISH STRATEGY IN THE NEAR FUTURE.

Report by the Chiefs of Staff Committee.

We have reviewed our Report on "British Strategy in
a Certain Eventuality" (Paper No. W.P. (40) 168) in the
light of the following Terms of Reference remitted to us
by the Prime Minister.

"In the event of France being unable to continue in
the war and becoming neutral with the Germans holding
their present position and the Belgian army being forced to
capitulate after assisting the British Expeditionary Force
to reach the coast; in the event of terms being offered
to Britain which would place her entirely at the mercy
of Germany through disarmament, cession of naval bases in
the Orkneys etc; what are the prospects of our continuing
the war alone against Germany and probably Italy. Can
the Navy and the Air Force hold out reasonable hopes of
preventing serious invasion, and could the forces gathered
in this Island cope with raids from the air involving
detachments not greater than 10,000 men; it being observed
that a prolongation of British resistance might be very
dangerous for Germany engaged in holding down the greater
part of Europe."

2.   Our conclusions are contained in the following
paragraphs.

331

3.   While our air force is in being, our Navy and
air force together should be able to prevent Germany
carrying out a serious seaborne invasion of this
country.

7. Germany could not gain complete air superiority unless she could knock out our air force, and the aircraft industries, some vital portions of which are concentrated at Coventry and Birmingham.

332

9. Whether the attacks succeed in eliminating the aircraft industry depends not only on the material damage by bombs but on the moral effect on the workpeople and their determination to carry on in the face of wholesale havoc and destruction.

10. If therefore the enemy presses home night attacks on our aircraft industry, he is likely to achieve such material and moral damage within the industrial area concerned as to bring all work to a standstill.

333

11. It must be remembered that numerically the Germans have a superiority of four to one. Moreover, the German aircraft factories are well dispersed and relatively inaccessible.

12. On the other hand, so long as we have a counter-offensive bomber force, we can carry out similar attacks on German industrial centres and by moral and material effect bring a proportion of them to a stand-still.

13. To sum up, our conclusion is that prima facie Germany has most of the cards; but the real test is whether the morale of our fighting personnel and civil population will counter balance the numerical and material advantages which Germany enjoys. We believe it will.

J.G. DILL.   (Signed)   C.L.N. NEWALL.
T.S.V. PHILLIPS.   DUDLEY POUND.
R.E.C. PEIRSE.   EDMUND IRONSIDE.

convenient moment; or, even worse, it would be another shameful Munich Agreement. Hoare hits back that rhetoric and bluster alone will not stop German panzers advancing down Whitehall, and that is what Britain now faces: total ruin and an end to a thousand years of freedom. He also argues forcefully that while the deal probably would not hold, it would at least buy time for Britain.

Halifax faces a crucial and momentous decision. His Cabinet is deeply divided. Hoare, backed by John Simon, Chancellor of the Exchequer, offers an apparently logical and rational way out of this calamitous war, albeit with a short-term fix. Churchill, backed by Anthony Eden, Secretary of State for War, and possibly even Archie Sinclair, the leader of the Liberals, reject the Hoare–Ciano Plan as a humiliating defeat that will solve nothing and only store up greater problems for the future. Chamberlain looks likely to sit on the fence. Halifax himself had criticised the Munich Agreement, but he is also a realist. What are his options? Either way, his decision could prompt a series of resignations and destabilise the government.

You look around the room as murmurs of discontent rumble on; it is up to you, as the leader holding the middle ground, to offer advice to Halifax. On your shoulders perhaps hangs the future of Britain.

AIDE-MEMOIRE

* Britain's troops on the continent are in great peril. Defeat seems imminent and if the troops are lost, the war is surely over.

* The Germans may not be able to defeat Britain, however. Consider the Chiefs of Staff paper tabled by Churchill (see Document 2). How realistic is this assessment?

* Time is running out. The Hoare-Ciano Plan offers a way to salvage something out of

the deteriorating situation before all is lost.

* Is the loss of territories in the Mediterranean and German occupation of the Channel coast too high a price to pay?

* A deal now might buy time to build for the future.

* Sam Hoare has a history of trying to secure agreements in times of crisis, most obviously the notorious Hoare-Laval Pact of 1935, which tried to surrender Abyssinia to Italy to appease Mussolini. That solved nothing. Of Hoare's part in it, one commentator stated, 'Gentlemen do not behave in such a way.'

* Churchill and his followers offer little other than bluster and rhetoric. It is surely the moment for reality to take over.

THE DECISION

Should you support the Hoare–Ciano Pact or not?

 **Yes** to the Hoare–Ciano Pact, go to **Section 4** (p. 44).

OR

 **No** to the Hoare–Ciano Pact, go to **Section 5** (p. 50).

SECTION 4
THE TREATY OF ROME

28 May 1940, Palazzo Venezia, Mussolini's headquarters, Rome

The atmosphere is tense in the Globe Room of Mussolini's Fascist government headquarters in Rome. Many of the leaders and officials of Europe's great powers are present, whispering and watching as Lord Halifax, Prime Minister of the United Kingdom, and Marshal Pétain, President of France, sign the Treaty of Rome, an agreement that will bring to an end the war that has raged across Europe since September 1939. On one side a smiling Adolf Hitler looks on, offering a few snatched comments to his Foreign Minister, Joachim von Ribbentrop, while on the other Il Duce, Benito Mussolini, nods knowingly and looks at his nephew, Count Galeazzo Ciano, Italy's Foreign Minister. For Mussolini this is a major diplomatic coup – he is portraying himself as the honest broker, the leading statesman who has saved the world from chaos and conflict.

You are the British Foreign Secretary, *Samuel Hoare*, and one of the key architects (along with Count Ciano) of the deal about to be signed. Highly experienced in British politics and government throughout the interwar years, you know this is your moment. To your mind, you have rescued Britain from defeat and total calamity and have headed off the blustering windbag populist Winston Churchill, a long-term rival of yours. Churchill's stubborn refusal to support Halifax and you in your negotiations almost brought down disaster upon the country. At the very least you have bought time for Britain – a pause in the war to allow a rebalancing and refocusing of priorities.

Yet as the world's press pho-
tographers snap away and the
film cameras capture the moment
of the countersigning by Hitler
and Mussolini, you are of course
uneasy. And when they are joined
by Halifax and Pétain on the bal-
cony overlooking Piazza Venezia
to wave to the crowds below to
signal the end of the war, you
are not naïve enough to believe
that confrontation with Germany,
and indeed Italy, is over. But
you and Halifax have brought
some realism to Britain's foreign affairs and have won some
breathing space.

6. Samuel Hoare

The Treaty of Rome is,
however, a severe setback for
the Western powers and a major
triumph for the Nazis and Fascists; it marks a clear victory for the Fascist military dictatorships of Europe, and a humiliating climb-down by
Britain. France has been rendered near-impotent and divided and is
now ruled by Marshal Pétain and Pierre Laval, who are far more sympathetic to Germany. Hitler's troops lord it across the Low Countries,
much of Scandinavia, Poland and now northern and western France.
Even Italy, without firing a shot, has secured major concessions in the
Mediterranean at the expense of the British and the French.

This is also a treaty that almost did not come to fruition. The final
Treaty of Rome is based heavily upon the Hoare–Ciano negotiations
that had caused such ructions in the British War Cabinet a few days
earlier. Halifax's decision to open up discussions with Hitler through
Mussolini had been backed by you, Chamberlain, John Simon and
even Clement Attlee, though he had to be nudged into line by Herbert
Morrison, his deputy. Churchill and Eden, however, were incensed
and an almighty row had broken out. Had nothing been learned from

Munich, Churchill raged? He would not be shamed by association with this new agreement. Halifax desperately tried to keep Churchill on board – the greater good, the needs of the nation. But Churchill could not be quelled; he resigned immediately and walked out of the Cabinet, taking Anthony Eden with him. The following day saw the Liberal leader Archie Sinclair and the First Lord of the Admiralty, the Labour politician A. V. Alexander, resign too. Halifax's government hung by a thread, but he and Chamberlain retained the support of the Conservative Party. Labour had split over the deal; some followed Attlee's lead in accepting the hard logic of negotiating a settlement, given the parlous state of Britain's defences, but others – led by Arthur Greenwood and Hugh Dalton – could not and resigned from Attlee's frontbench team.

Halifax instructed you to pursue further talks with the Italians and, after a period of frantic negotiation, a ceasefire was agreed. Hitler's demands were hugely damaging and of course humiliating, but the new government in Paris saw no alternative but to agree; London was therefore left with no realistic option but to comply. The loss of Malta to the Italians was the main price Britain had to pay Mussolini for opening up the route to a deal, although it could have been much worse. There was great despair back in Britain over the terms of the deal; nonetheless, after a rancorous debate, the Treaty of Rome was ratified by the House of Commons in early June and the war came to an end.

AFTERMATH

Churchill spoke out repeatedly against the shame of Rome, but his retreat to the back benches hindered his efforts; he returned to more years in the wilderness, retirement to Chartwell and writing his bitter memoirs. In the late summer of 1940 Halifax called a general election and although some losses were suffered to the Labour Party, the pro-treaty Conservatives won the day easily enough and secured a comfortable majority; the public was not minded to reopen a war that Britain had so obviously been losing quite badly.

The wider and longer-term implications were profound for both Britain and the world. Dismayed at the feeble efforts of Western Europe to contain Hitler, Roosevelt's administration retreated from European affairs still further, seeking security in hemispheric defence and a robust stance against Japanese militarism. America's future now seemed focused squarely on the Asia-Pacific region. For Japan, the consequences of Hitler's success in Europe were significant. Though France and Britain were humbled, they were not defeated to the extent that they could not react to Japanese expansion in South-East Asia. With the USA stepping back from European affairs, Japan's prospects of breaking the impasse of the stalemated war they had been fighting against China since 1937 appeared remote.

In Eastern Europe, Germany's success bolstered still further growing animosity towards the Soviet Union. In the summer of 1940 Stalin's state appeared hopelessly exposed; any lingering belief he had that the West might keep Hitler at bay had evaporated, and it was patently clear that Hitler had his eyes fixed on expansion to the east. Stalin's efforts to shore up his position by demanding territory from Finland, the Baltic States and Romania had merely hardened attitudes against him. Hitler's diplomatic efforts to forge an anti-Soviet alliance in the Balkans proved fruitful. British intelligence efforts to prevent the formation of the alliance came to nothing; there was little trust now in anything London said.

Germany's plans for a war in the east developed rapidly. Halifax's deal with Germany had avoided any ruinously costly air campaigns over Western Europe in the summer of 1940, and German military preparations were equally spared the efforts of a maritime war against Britain. Everything was focused on the USSR. In the Mediterranean, conflict exploded in Egypt in the spring of 1941, when British forces had to confront a rebellion led by Egyptian resistance fighters and revolutionaries, which many believed had been backed by the Italians. Now even Britain's Middle Eastern power base began to crumble.

In late April 1941 Hitler launched an all-out offensive against the Soviet Union, a war that would determine the future history of Europe, and possibly the world. With France emasculated, Britain neutral and weakened, and the USA looking at the Asia-Pacific region, few held out any hope for Stalin's Soviet Union. An alternative regime in London might have looked to exploit the situation, but Halifax's government had little stomach for further hostilities. The die had been cast.

THE DECISION

▶ To explore other alternatives, go to **Section 5** (p. 50), or to **Section 2** (p. 24) to examine what might have happened if Churchill had become Prime Minister.

OR

▶ To explore the history of these events, go to the **Historical Note** on p. 65.

SECTION 5
HALIFAX'S WAR

5 June 1940, Cabinet War Rooms below Whitehall, London

A tense meeting of Lord Halifax's War Cabinet is under way and the friction and hostility in the room are apparent for all to see. From the moment Sam Hoare's proposal to open negotiations with the Italians for a possible deal or ceasefire was scotched at a rancorous and turbulent Cabinet meeting a little over a week earlier, there has been simmering hostility between the Churchill and Hoare factions in the government. You are **Anthony Eden**, the dashing and highly regarded forty-two-year-old Secretary of State for War in Halifax's government. You had previously served as Foreign Secretary in the governments of Baldwin and Chamberlain in the 1930s, but had resigned over appeasement policies. When Halifax had formed a new government in the wake of Chamberlain's resignation, you were brought into the War Cabinet, with Churchill's backing. You had reservations about Winston's abilities, but over recent weeks you have come to respect his drive, determination and grip. He has been acting as Defence Secretary, effectively running the war under Halifax's oversight. Yet Lord Halifax, despite being Prime Minister, seems a secondary figure in the war effort – more so since Hoare's plan foundered. In contrast, Churchill's direct involvement in the organisation of the successful Dunkirk evacuation that saved 300,000 Allied troops has drawn plaudits and popular respect, even though you have also heard from your military leaders that Winston's meddling is not always welcome.

With Churchill's stock at a new height, he is bombarding the Cabinet with proposals and ideas, but more than anything he is lording it over the beleaguered Foreign Secretary, Sam Hoare. His opponent's defeatism has been exposed, Churchill glowers: the British Army has been saved, the RAF and the Royal Navy are still fighting and so even if France succumbs, there is no need to make deals with the Nazis. To have even considered it was a colossal misjudgement on Hoare's part, Churchill proclaims. Chamberlain and John Simon, allies of Hoare, can do little to deflect the attack; and the neutrals in the Cabi-

7. Anthony Eden

net, such as Attlee, say little, because deep down they believe Churchill is probably right. Hoare is compromised again, as he had been back in the mid-1930s over the deal that he cooked up with Pierre Laval concerning Italy's invasion of Ethiopia, something that forced Hoare's temporary resignation from the government.

Halifax had initially defended Hoare over his secret negotiations with the Italians the previous week, but was very uneasy about the matter. For him, it had been one thing to negotiate to keep Italy out of the war to try and keep France fighting; it was quite another to get into what were obviously direct dealings with Hitler. He had been greatly troubled by Chamberlain's Munich Agreement in 1938 and clearly did not think it wise to follow the same path now. And Halifax can see that support for Hoare is waning fast. You watch on as the criticism mounts. Secretly you hope that Halifax may turn to you to take on the role of Foreign Secretary if Hoare resigns; after all, you have previously held the post. Churchill later confides that he sees you as the natural replacement for Hoare.

But Halifax is not known as the Holy Fox without good reason. Conscious of trying to maintain a balance between the factions in the

War Cabinet, he surprises everyone the following day when he announces that, as Hoare has accepted the inevitable and tendered his resignation, Rab Butler will be taking up the post of Foreign Secretary. Churchill and you fume; Butler is known as having been a supporter of appeasement, and he is also 'a good party man', one who will toe the line and support Halifax, if needs be. Commitment to continuing the war whatever happens is therefore still not securely embedded in the Cabinet. And Britain's position is still perilous. In the following days Belgium surrenders and the French Prime Minister Paul Reynaud hints that he is losing control in Paris to a group of 'defeatists' who seem determined to get France out of the war, come what may. The issue is still in doubt.

AFTERMATH

Over the next few weeks France was eventually forced to surrender and was partitioned, with the rump state being headed by a collaborationist regime under Marshal Pétain and Pierre Laval at Vichy. Britain faced disaster, and the optimistic words of the Chiefs of Staff (from their future strategy assessment, see Document 2 on p. 39) – that if the RAF could prevent German domination in the air as much as the Royal Navy would at sea, then invasion could be averted – were to be sorely tested. In these trying times it soon became clear that Halifax was no Churchill; a clear, calm, erudite, managerial approach to leading the United Kingdom during this perilous period was not what the nation needed. Pugnacity, single-minded focus and determination, alongside sheer resolve, were of more use. Of course the resolution of the fighting forces and of the nation's population were to be the decisive factors in ensuring survival, but leadership in these times was crucial, too. As Defence Minister, Churchill was in effect running the war, with a scrutinising overview provided by Halifax and Attlee.

Matters came to a head in the spring of 1941. Having survived the onslaught of the German air forces and U-boats through the dark times of the Battle of Britain, the Blitz and the Battle of the Atlantic, Britain was in a stable position by March 1941. Hitler had focused his attentions elsewhere and Britain's Mediterranean campaign against Italy was going well enough, though there was still no clear route to victory.

Halifax, prompted by his new Foreign Secretary, Rab Butler – and without open discussion in the War Cabinet – decided upon a course of action to test the waters with Germany. Butler used contacts with the Swedish embassy in London to try and ascertain what Hitler would be willing to negotiate away to ensure Britain's neutrality. If Hitler was intent on attacking the USSR later in 1941, as seemed possible, he could be willing to come to favourable terms to get the British out of the war.

8. Rab Butler

Churchill and Eden were tipped off, provoking an almighty row in the War Cabinet. Butler argued that putting out feelers from a position of some strength was quite different from the chaos of the previous year.* If Germany attacked and crushed the USSR, as seemed likely, Hitler would have no need to offer favourable terms to Britain, but he might do so before such an invasion to focus all his resources eastwards. Now was the moment to see if a favourable deal was possible. Churchill fumed that such a craven approach to the Nazis suggested weakness, when he and the armed forces had demonstrated that Hitler could not defeat Britain.

This time Churchill and Eden, with Attlee and Sinclair on board too, called for Butler to resign and questioned Halifax's commitment

---

* Butler was later regarded as having been a true believer in appeasement and he did speak to the Swedes in 1940 during the May crisis, something that very nearly cost him his job.

to continuing the war. Halifax's position was far less secure by March 1941; Hoare had gone, Chamberlain had died and now Butler was in the firing line. Facing humiliation, Halifax resigned, taking Butler with him, and Eden was called upon once more to take up the post of Foreign Secretary. When Halifax departed, Churchill was the only possible replacement and he therefore became Prime Minister a month before Hitler invaded the USSR – the moment when the grand strategic direction of the war began to shift.

THE DECISION

▶ To explore other alternatives, go to **Section 4** (p. 44), or to **Section 2** (p. 24) to examine what might have happened if Churchill had become Prime Minister.

OR

▶ To explore the history of these events, go to the **Historical Note** on p. 65.

SECTION 6
HALIFAX'S GAMBLE

30 May 1940, Rome

A cavalcade of limousines glides through the streets of Rome, protected by a rather ostentatious and somewhat intimidating military escort. The cars carry a British delegation of officials and diplomats headed by Lord Halifax, the Foreign Secretary. In one of the limousines you sit alongside Halifax, because you are ***Alexander Cadogan***, Permanent Undersecretary at the Foreign Office, the department's chief civil servant. You have held the post since 1938 and had been previously intimately involved in the Munich Agreement – something over which you had deep reservations. You are a well-connected and distinguished diplomat from a wealthy family, an Old Etonian and a graduate of Balliol College, Oxford. As you arrive in the Piazza Venezia, the centre of Benito Mussolini's Fascist government in Italy, you have grave concerns. Halifax is here to explore the possibility of negotiating a settlement to the current war with Germany; the Italians are acting as honest

9. Alexander Cadogan

brokers, though they clearly favour the German side and are out to extract concessions in the Mediterranean from Britain and France as the price for their involvement in the talks.

Halifax has confided to you that he is trying to limit the damage to the British Empire caused by Churchill's obdurate attitude to continuing the war, whatever the cost. Halifax had believed, when he acquiesced to Churchill becoming Prime Minster ahead of him on 9 May, that Churchill's flowery rhetoric and mercurial manner might well hinder the conduct of the war. In Cabinet on 26 May Halifax witnessed exactly what he had feared: Churchill's foolhardy bulldog determination becoming a liability. Despite Halifax's protestations in Cabinet, Churchill had blocked his attempts to find out what the Italians might want in order to stay out of the war, or even to explore what an overall settlement with Germany might look like. Nothing had to be accepted, and Halifax certainly did not want a repeat of Munich, but it was surely logical and sensible for all avenues to be pursued to maintain Britain's independence and security? Indeed, it was incumbent on Halifax, as Foreign Secretary, to do just that, although Churchill seemed oblivious to the realities of the situation.

The following day, however, Churchill was confronted by a deteriorating military situation on the continent, Belgium's surrender and whispers of France trying to come to some form of agreement with Mussolini. At the first War Cabinet meeting of the day Halifax pressed hard to be mandated to follow up on his initiative with the Italian ambassador – the Italians seemed to be willing to act as intermediaries with Hitler. Halifax was unsure about the degree to which this should be pursued, but all options should surely be explored? Churchill hesitated and tried to talk Halifax round in a private conversation in the gardens at Number 10, but Halifax was not to be persuaded, or so he later told you. It seems that Churchill had been worried Halifax might resign and take Chamberlain with him, so finally he had grudgingly acquiesced to Halifax's initiative. You heard on the grapevine that Churchill had brought Archie Sinclair – the leader of the Liberal Party, and a known friend and colleague of the Prime Minister – into the discussions the following day to bolster his support against Halifax and Chamberlain, but by then it was too late.

On Tuesday 28 May Halifax had returned to Cabinet with a set of draft proposals from the Italian ambassador, Giuseppe Bastianini. It seemed that Mussolini was willing to stay out of the war and act as an honest broker between the Allies and Hitler, in return for Malta and Gibraltar and the withdrawal of British forces from Egypt, other than on the Suez Canal. Halifax claimed that it was more than possible to save Gibraltar in negotiations. What mattered now was the German pound of flesh. During the meeting you arrived from the Foreign Office with the latest update from Paris: Paul Reynaud, the French Prime Minister, was struggling to keep his government together and a move to open negotiations with Hitler, regardless of the alliance with Britain, was being openly discussed there. In the confusion French military resistance was collapsing around Dunkirk, exposing the British forces trying to evacuate across the English Channel to extreme German military pressure. There was little likelihood of a miracle now.

Churchill was cornered; he had argued that the terms available to Britain would be so humiliating and crippling that they would never be acceptable, but with the military position deteriorating and France on the point of collapse, this offer from the Italians was almost worth thinking about, especially if the Germans could be persuaded to allow the British Army on the continent to be withdrawn safely, and France could be preserved in some way. Chamberlain supported Halifax's desire to pursue the negotiations; Attlee and even Sinclair hedged. Churchill risked ripping his government to pieces if he blocked Halifax's initiative.

Matters deteriorated quickly. Even as Halifax began preparations, the very fact that the British were contemplating talking to the Italians leaked and chaos erupted in Paris. Reynaud was forced to resign and Pétain took control, immediately seeking talks with the Germans, directed by Pierre Laval, the new French Foreign Minister. Churchill's efforts to rally support at home had also failed. Although the British military Chiefs of Staff had argued a few days earlier that Britain might be able to fight on without France (recall Document 2 on p. 39), their resolution now started to crumble, too. Churchill let it be known that he regretted not replacing the floundering General 'Tiny' Ironside

10. The Fascist Grand Council Room in Rome, scene of the negotiations

as head of the army a few days earlier with the more ebullient General John ('Jack') Dill, who might have offered more positive support. There were few options left, and Churchill had acquiesced to Halifax flying to Rome to meet Laval and Joachim von Ribbentrop, the German Foreign Minister, to open tentative talks.

As you arrive and walk into Mussolini's Grand Council Room to open negotiations, gloom descends. The meeting is chaired by Count Ciano, the Italian Foreign Minister, but the German terms are crippling for France: partition, demilitarisation, loss of Alsace-Lorraine and Corsica, and German occupation of Paris and the English Channel and Atlantic coastlines. Britain will have to return the colonies taken from Germany at the end of the Great War, agree to a preferential Empire trade deal with the Third Reich and hand over territories to Italy.

A deflated Laval acquiesces but, despite the almost unbearable pressure in the room, you persuade Halifax that the matter will have to be taken back to London. Ribbentrop is furious and Ciano tries to apply pressure, arguing that the deal will have to be accepted now or risk being withdrawn. To you, however, the deal is unacceptable and even Halifax hesitates – it will have to be discussed by the British government back in London. You return to Britain with all haste and outline to the War Cabinet the deal being offered.

The meeting erupts in fury and recriminations. You and the other officials look on as the politicians clash. Churchill refuses to agree to the terms, and so an exasperated Halifax resigns in frustration, but Chamberlain, perhaps conscious of his Munich legacy, hesitates and the government survives. Nonetheless, the very act of tentatively opening discussions has proved calamitous; France has quit the war, the British Army on the continent is being battered by the German army with little prospect of any respite, and Churchill's credibility has been severely dented.

AFTERMATH

The war carried on through the summer, but Churchill's power base had been compromised and the British Army crippled. When German air attacks intensified in August, and with the German U-boats operating from midsummer out of newly acquired French Atlantic ports threatening to strangle vital shipping, Britain's survival looked highly unlikely. The possibility of a direct invasion loomed large and in a desperate effort to head off total calamity, Lord Halifax, backed by senior Tories and other interested establishment groups, engineered a parliamentary coup against Churchill. Halifax returned to high office, this time as Prime Minister and with a clear policy of ending the war with Germany. Hitler had been persuaded by his military commanders that mounting an invasion of Britain was almost impossible in the autumn of 1940, and his attention was in any case now fixed squarely on attacking the USSR, which he was planning for 1941. With Hitler keen to bring the war with Britain to an end, the negotiated settlement was ultimately similar to that offered in Rome a few months earlier.

Germany's preparations for the invasion of the USSR intensified and, now spared a damaging simultaneous conflict with the British Empire, Hitler's chances of success looked promising. With Paris and London neutralised, and a despairing, isolationist USA looking elsewhere, Stalin's hopes of survival against a Nazi onslaught appeared remote indeed.

THE DECISION

▶ To explore other alternatives, go to **Section 7** (p. 62), or to **Section 3** (p. 34) to examine what might have happened if Halifax had become Prime Minister.

OR

▶ To explore the history of these events, go to the **Historical Note** on p. 65.

SECTION 7
CHURCHILL'S HOUR

27 May 1940, Cabinet War Rooms, Whitehall

As the War Cabinet assembles and begins discussing the strategic situation, Churchill initiates a carefully crafted game. He needs to enforce the policy of continuing the war even if France falls, and therefore the British government must not send out any formal signals at all to the enemy about possible terms of settlement. Yet it is also essential that Churchill keeps both Halifax and Chamberlain in his government; if they resign now they could take much of the Tory Party's support with them. So if Churchill refuses point-blank anything that Halifax suggests, his Foreign Secretary could walk away and the government might fall.

But Churchill has bolstered his support on the Defence Committee of the War Cabinet, the key decision-making organ in this process. And it is you, as an old friend and colleague, that he will be relying upon; you are *Archibald 'Archie' Sinclair*, Secretary of State for Air – in effect the political chief of the RAF. As an Old Etonian, ex-Guardsman and one of the largest landowners in Britain, you are powerful and influential, as well as being irresistibly charming, handsome and very well regarded. You have been a strong voice against appeasement and for standing firm against the Nazis for many years, but more importantly at this time you are also leader of the Liberal Party. Winston has brought you onto the Defence Committee ostensibly as a representative of a major party, but in truth he needs you as an ally. He has also confided that he believes Attlee and Greenwood of the Labour Party are likely to back him, too – and that should give you a majority, if push comes to shove.

The meeting begins with Halifax pressing his case for opening tentative discussions with the Italians. Churchill stalls, bolstered by his belief that he can overcome the doubters in the room and isolate Halifax. Churchill argues that he does not believe there are any circumstances in which Britain would get an acceptable deal from Hitler, so there was little point in trying; it would merely signal weakness to the enemy. Halifax is clearly annoyed. This seems to be a hardening of the Prime Minister's

11. Archibald Sinclair

position. Are there no terms at all on which Churchill might seek a negotiated settlement? Stalemate.

Discussions continued throughout the day in between meetings, and you also hear that Halifax and Churchill held a private chat in the gardens at Number 10.* It is possible that Churchill tried to persuade Halifax that resignations would not help the country at all in the prevailing crisis.

28 May 1940, Cabinet War Rooms, Whitehall

Churchill may have had some success with Halifax, but it is not compelling enough to prevent Halifax returning to the subject of discussions with the Italians the following day. He argues that Britain might get better terms by negotiating before France surrenders rather than after, and as France is in turmoil, the British government should at least

---

* No records were kept of this chat, so we do not know for sure what passed between them.

explore what the lifeline potentially being offered by Italy might entail. Arthur Greenwood argues that this is not the time to talk about capitulation – a term that riles Halifax. Interestingly, you notice that Chamberlain does not come to Halifax's aid, perhaps sensing that the War Cabinet is minded to back Churchill. Churchill stalls again, not wanting to infuriate Halifax still further, and then plays his trump card.

In between War Cabinet meetings he speaks to the rest of the full Cabinet, beyond the small Defence Committee, to bolster his support. He states that Britain will have to fight on, come what may, and that there is nothing to be gained by talking to Hitler. He goes on: 'If this long island story of ours is to end at last, let it end only when each one of us lies choking in his own blood upon the ground!' Churchill receives the Cabinet's full support. Finally realising that he was in minority of one, Halifax backs down and, in his usual manner of accepting duty when called upon, does not resign. Churchill has carried the day and the war will go on regardless.

AFTERMATH

Churchill's position was cemented by the near-miraculous escape of the huge numbers of Allied troops from Dunkirk, and then by the failure of the Germans to win the Battle of Britain. His bullish refusal to countenance any idea of dealing with Hitler shored up support for his government and secured his grip on power. By the autumn Chamberlain had succumbed to cancer and Halifax had been moved to Washington as ambassador to the USA, an important though less prestigious role.

Britain had survived the crisis of 1940, but until the entry of the USSR and the USA into the war the following year there was no obvious route to victory, other than wait and see. This was not in itself wholly unrealistic, given the inherent contradictions and weaknesses of the Third Reich, but as Churchill secured his position as Prime Minister in the summer of 1940 it was by no means obvious that Britain had chosen the right path, despite the way events subsequently unfolded.

THE DECISION

▶ To explore other alternatives, go to **Section 6** (p. 56), or to **Section 3** (p. 34) to examine what might have happened if Halifax had become Prime Minister.

OR

▶ To explore the history of these events, see the **Historical Note** below.

HISTORICAL NOTE

After the passage of time and the events of the Second World War it seems perfectly obvious and clear that Winston Churchill was the man of the hour, but on 9 May 1940 that was most certainly not the case. The historical path that takes you from Section 1 to Section 2 and then to Section 7 was by no means the most obvious route, and even then Churchill's position was by no means secure – he was heavily reliant on the support of Chamberlain and Halifax, who were watching him like hawks, such was their unease at Churchill attaining power.

If you decided to explore what might have happened if Halifax rather than Churchill succeeded Chamberlain (from Section 1, go to Section 3), you would have followed what for many people in May 1940 seemed the obvious path. Ironically, considering how the party has since adopted him as their all-time iconic leader, and how modern-day Tories always hark back to the Churchillian spirit, many Conservatives in 1940 were horrified that such an untrustworthy rogue as Churchill might become their Prime Minister. Ultimately, if Halifax had wanted to become Prime Minister, it would have happened – he had the backing of the Conservative Party, and Attlee and Sinclair would have accepted him as leader.

Churchill, however, was determined and seized his opportunity – the infamous 'silence' when Chamberlain asked what his position was, regarding Halifax taking over (Section 1 to Section 2). Subsequently Churchill's careful manoeuvrings also enabled him to head off Halifax's initiative of talking to the Italians, and this is again the historical path that takes you from Section 2 to Section 7. It is also a route that is often described as one based on a bulldog spirit, born of quixotic determination rather than hard logic – a strategy of the heart more than the head. This does Churchill, and indeed those who supported his stance in Cabinet in May 1940, too little credit. Churchill believed it was pointless negotiating with Hitler because any agreement would be near-worthless, as Hitler clearly could not be trusted. In addition, any deal would be crippling and would seriously compromise any chance of taking up the struggle again in the future, while any hint of serious negotiation was also likely to display a lack of resolve and push Britain down the slippery slope to another Munich. Churchill, backed by considered military assessments, believed that even if France was forced out of the war, defeat for Britain in 1940 was some way off. And in the long term Britain, in conjunction with likely future Allies, still held a number winning cards. Ultimately Churchill's strategy in the summer of 1940 of continuing the war, come what may, was based on cold logic as much as rhetoric and hope.

It should also be noted that Halifax's option of opening up very tentative discussions with Italy, and his then forcing through the idea against Churchill's better judgement (going from Section 2 to Section 6), was neither completely foolish nor weak. Indeed, it was Halifax's job as Foreign Secretary to explore all these avenues and provide such alternatives to the Cabinet in May 1940. Although the Axis powers were clearly in the ascendancy by late May 1940, victory against France was by no means assured, nor would it be won easily; it remains a myth to this day in the anglophone world that the French tamely threw in the towel in 1940, when in fact they battled hard, suffering heavy loss of life (in excess of 90,000 French soldiers were killed fighting the Nazis in the 1940 campaign). This costly and attritional fighting continued for many weeks after Dunkirk – fighting that ultimately severely damaged Germany's war effort. It is not without foundation then that Germany might have been willing to avoid these ongoing losses if a deal with both Britain and France could be negotiated.

It is unlikely that, as Prime Minister, Churchill would have accepted any deal, even if he had been forced down that path by Halifax and Chamberlain. But what if Halifax had been Prime Minister (from Section 1, go to Section 3)? He was not a man directed by such a self-assured grand vision as Churchill; he perceived himself as a managerial, transactional 'realist' and, as such, negotiations with the Axis powers in late May / early June 1940 were more likely. The path from Halifax becoming Prime Minister and then exploring options with the Italians (Section 3 to Section 4 or 5) is therefore distinctly possible, but it is far from clear what Halifax might have done in response. We can reliably estimate Churchill's response, and of course much would have depended on who inhabited Halifax's War Cabinet and how it was constituted – historically Churchill cleared out 'appeasers' such as John Simon and Sam Hoare, but Halifax was unlikely to do so, though Attlee and the Labour Party might have pressed him on this. Additionally, when Churchill became Prime Minister he reshaped the whole structure and created a small Defence Committee run by him, to grip the war. Halifax was more likely to keep a larger War Cabinet in a similar fashion to that of Chamberlain, and thus the chance of rancour and disagreement was much greater.

Yet although it is possible that Sam Hoare, a long-term opponent and rival of Churchill, might have brought forward a deal such as that discussed and passed in the Treaty of Rome (Section 4), it is more likely that Halifax, having previously been sceptical about Munich, would not have accepted (Section 5). But whether he was then the right man for the job of leading the United Kingdom is another matter. Perhaps his own self-awareness, which had prevented him seizing the opportunity in May 1940, would ultimately have been his undoing as Prime Minister. Halifax was far from the weak and supine character popularly portrayed in films such as *Darkest Hour* (2017), but – like Chamberlain before him – he was probably not well suited to commanding and driving forward a war effort such as that demanded by the Second World War.

In the twenty-first century political figures such as Churchill have come in for more forensic scrutiny and there is now greater awareness of what most would regard as his less salubrious side – racist rhetoric and actions, strong pro-imperial attitudes and a range of clearly questionable political decisions. But ultimately in the period of 1940–42 Churchill's grip, determination and unwavering grand strategic vision were crucial factors in driving forward Britain's war effort at a time when the issue was most certainly in doubt.

What of the other roles in this scenario? Chief Whips are often key players during periods of succession in political parties, and **David Margesson** was no different. His input was crucial in determining who should succeed Chamberlain. Indeed, Halifax later noted that he thought Margesson leaned towards Churchill on 9 May as the next Prime Minister. Margesson's role was all the more remarkable because at the time he was going through a divorce and was living at the Carlton Club, as he had no home. Later, when the club was bombed, he slept on a camp bed in the Cabinet War Rooms. Margesson was generally well liked and popular, despite being Chief Whip, and Churchill first retained and then promoted him to Secretary of State for War in October 1940, a post in which he proved quite competent. Nevertheless, when Britain suffered a series of military setbacks in early 1942, Margesson accepted that he had to be replaced.

Hastings 'Pug' Ismay acted as Churchill's chief military assistant throughout the Second World War and was someone the Prime Minister relied upon for advice and insight. Ismay was often a moderating influence on Churchill and acted to maintain good working relations between the military service and the PM. He later travelled with Churchill to many of the key international conferences, where he maintained effective relations with the Americans, too.

Clement Attlee, despite the teasing comments made by Churchill about his demeanour and temperament, remained supportive of the PM throughout the war. Though quiet and moderate, his support for Churchill against Halifax and Chamberlain in May 1940 was crucial. In 1945 Attlee led the Labour Party to a landslide victory in the general election against Churchill's Conservatives and changed Britain for ever.

Samuel Hoare locked horns with Winston Churchill for many years in the 1930s, so when Winston became Prime Minister in May 1940 Hoare's time in high office effectively came to an end. He was always identified as an arch appeaser, although even his patience ran out by the time of the invasion of Poland. From 1940 to 1944 he served as ambassador to Franco's Spain.

Anthony Eden served under Churchill during the war, mainly as Foreign Secretary, and many foresaw that he would replace him as leader of the Conservatives when Churchill eventually stood down. But despite the crushing defeat to Labour in 1945, Churchill carried on, and Eden had to wait until 1955 before he got the top job. It did not go so well and he was undone by the Suez Crisis.

Alexander Cadogan served as Permanent Under-Secretary at the Foreign Office from 1938 until 1946 and was party to the major diplomatic decisions and wrangles throughout the war. He was particularly dismayed by the decisions at the Yalta Conference in 1945, which effectively sealed the fate of smaller powers such as Poland at the hands of the USSR. After the war he served as the UK's permanent representative to the United Nations (1946–50) and as chairman of the governors of the BBC (1952–7), even though he had never seen a BBC programme at the time of his appointment.

Archie Sinclair served successfully as Secretary of State for Air in Churchill's wartime government from 1940 to 1945. He always cut a dashing and popular figure, but in leading the rapidly declining Liberal Party, his frontline political career effectively came to an end in 1945 when the Liberals were reduced to just twelve seats, of which Sinclair's was not one. Until Jo Swinson's defeat in 2019 he was the last major party leader to lose their seat at a general election.

2

THE MEDITERRANEAN GAMBLE:

THE WAR IN NORTH AFRICA, 1941–42

SECTION 1
O'CONNOR'S BATTLE

7 February 1941, Lieutenant General Richard O'Connor's tactical headquarters, Ghenimes, Libya

It has been a day of stunning success for the British Army's Western Desert Force in Libya. The Via Balbia, the main highway along the Libyan coastline built on the orders of the Italian Fascist dictator Benito Mussolini, is jammed with convoys of vehicles carrying captured

1. Brigadier Eric Dorman-Smith

Italian soldiers north to Benghazi and onwards to captivity. Littering the roadside are abandoned and destroyed lorries, armoured cars and tanks of the routed Italian Tenth Army. It has been a great victory for the Allies.

You are **Brigadier Eric 'Chink' Dorman-Smith**, a forty-five-year-old Irish officer in the British Army, working for General Archibald Wavell, the Commander-in-Chief of the Middle Eastern theatre.* You are known as something of an innovative thinker, mercurial and instinctive in nature, someone replete with ideas.

---

* The nickname relates to a comment that Dorman-Smith resembled a Chinkara antelope regimental mascot.

Indeed, you are one of the key architects of the plan that has resulted in this great success. Wavell, based back in Egypt, recently despatched you to the front line in Libya to report on progress and advise on the next steps.

You arrived to find the commander of the Western Desert Force, the dynamic and astute Lieutenant General Richard O'Connor, directing operations from a makeshift tactical headquarters. He has been constantly on the move, reacting to the ever-shifting position of the Axis and Allied armies sweeping across Libya. You have been impressed by the fifty-one-year-old O'Connor, who appears to be a 'modest, neat, small, compact but clear-minded person: wholly self-confident'. Never one for publicity or overt self-promotion, he nevertheless elicits considerable backing from his team, who clearly consider him a 'brilliant commander'.

Just that morning at 9 a.m., while breakfasting on cold sausage, you had been with O'Connor when he received news that the entire Italian Tenth Army had surrendered, following their decisive and crushing defeat at the Battle of Beda Fomm. Some 130,000 enemy troops were soon to be heading into captivity. As you and O'Connor now drive forward to the headquarters of 7th Armoured Division – the famous Desert Rats – to inspect the battle's outcome, the opportunities opening up seem potentially overwhelming. There is great pressure, for if the right decisions are made quickly, all of Mussolini's remaining forces in North Africa could be captured in just a few weeks.

It had all seemed so different back in the summer and autumn of 1940. Then, Mussolini's Italy had declared war on France and Britain on 10 June, hoping for easy pickings in the Mediterranean against the reeling and collapsing Allies. British and Italian forces soon began skirmishing along the Libyan-Egyptian border: Libya had been an Italian imperial possession since 1896, while Egypt, though not formally part of the British Empire, was to all intents and purposes under the control and direction of London. For Churchill's government it was imperative to hold Egypt and prevent Mussolini, and through him Hitler, gaining access to the Middle Eastern oil fields, which were also controlled by the British.

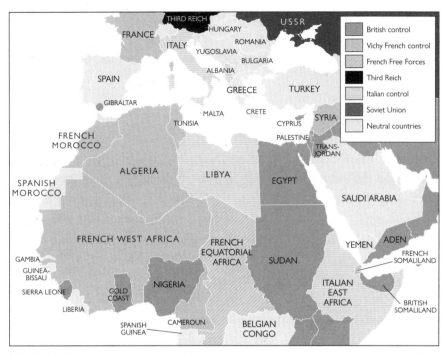

	British control
	Vichy French control
	French Free Forces
	Third Reich
	Italian control
	Soviet Union
	Neutral countries

Map 1: Mediterranean Theatre, 1940

Mussolini had assembled a sizeable army in Libya with more than 150,000 troops, 600 armoured vehicles, 1,600 pieces of artillery and in excess of 300 aircraft. In September, and despite relying heavily on a supply line stretching back almost 1,000 miles to Tripoli, his army began to advance into Egypt. The British forces on the border fell back, but the Italians only pushed some sixty miles into Egypt before supply issues and the need for reinforcements prompted a halt. They established five camps and dug in.

But the British commander of the Middle Eastern Theatre, General Archibald Wavell, was planning action of his own against the Italians. The fifty-seven-year-old Wavell was a studious, careful and taciturn man, the complete antithesis of his political boss in London, the ebullient and charismatic Winston Churchill. They were soon at loggerheads, with Wavell requesting extra troops and equipment, and Churchill in contrast demanding immediate results to shore up his position at home. Wavell had visited London in the late summer of 1940 but, despite his careful and (according to Anthony Eden) 'masterly' strategic assessment of the Middle East situation, Churchill was not convinced, stating that Wavell would merely make a 'good chairman of a local Tory association'. 'I do not feel in him that sense of mental vigour and resolve to overcome obstacles which is indispensable to successful war,' he grumbled. For his part, Wavell thought Churchill's military sense had got stuck in 1899. They never really understood each other at all.

Churchill had been further bemused by Wavell in October. The Italians had just invaded Greece and the British government began planning to send troops from the Middle Eastern Command to aid the Greeks, a move that greatly concerned Wavell. It was only then that he confessed to London that he was in fact planning an offensive against the Italians in Egypt, based partly on an assessment that you had produced. Although it was initially supposed to be a limited 'raid', Wavell had instructed his commanders – most notably O'Connor – to seize any opportunity to exploit the situation still further if such an eventuality presented itself.

2. O'Connor and Wavell

The difficulty, it seemed to you and Wavell, was that Churchill needed quick results to convince him that attacking the Italians in the deserts of North Africa was a better option than sending help to the Greeks, who were now embroiled in holding back an Italian invasion from Albania. The Prime Minister was conscious of the bigger picture and wanted a military success against the Italians to shore up pro-British support in beleaguered Greece, and to dissuade Turkey from becoming involved in the war. If Wavell could not deliver quick success, Churchill believed he would have to send troops direct to Greece to aid the Athens government.

When Wavell finally gave the go-ahead to O'Connor to launch the attack, much was hingeing on its success. Wavell remained cautious in his assessment of what was then codenamed Operation COMPASS: 'I am not entertaining extravagant hopes of this operation' but 'I do wish to make certain that if a big opportunity occurs we are prepared morally, mentally and administratively to use it to the fullest.' It was soon evident that such a big opportunity was about to present itself.

By the time COMPASS began, O'Connor's Western Desert Force had been boosted by new equipment, troops and resources, and although on paper Allied forces were still heavily outnumbered by the Italian Tenth Army (32,000 to *c.*150,000, by your estimates), your forces were much more mobile and were deployed to exploit this superiority against a largely passive and reactive opponent. The British Army had also recently introduced a battalion of Matilda II tanks, which, despite being slow and somewhat unreliable, had armour protection so thick that no Italian anti-tank gun appeared to be able to deal with them. Fused with a dynamic and aggressive plan based on manoeuvre and seizing and retaining the initiative, O'Connor believed great success was possible.

Within a few days of the attack beginning, Allied forces had unhinged the entire Italian position and the enemy began a disorderly retreat that soon became a rout. The Western Desert Force mopped up more than 38,000 enemy troops (including four generals), more than 230 guns, seventy-three tanks, many transport vehicles and a vast amount of supplies and stores (see Map 2).

The list of disasters that befell the Tenth Army grew as the Western Desert Force continued its advance at pace, capturing Bardia, Tobruk and Benghazi. At the Battle of Beda Fomm a final collapse of the Italian Army in Cyrenaica was confirmed. O'Connor's forces had driven the Italians right across Cyrenaica, advanced some 500 miles and eventually accounted for more than 140,000 Italian troops, some 400 tanks and in excess of 500 aircraft.

But what now? As you and O'Connor stop at Ghenimes, halfway between Beda Fomm and Benghazi, you consider your options. Allied advance reconnaissance forces have pressed on to capture Agedabia, and have even probed as far as El Agheila and on into Tripolitania. The road appears to be open. O'Connor believes that Italian power in North Africa could be extinguished by a sudden rapid thrust further up the coast to Sirte, some 160 miles further west, and then up the coastal road to Tripoli itself, the capital of Italian Libya. Your advance unit, the 11th Hussars, is straining at the leash, and you also have a brigade of the 6th Australian Division equipped and ready to move at a given signal. Other forces could soon follow up, if the logistical situation allows.

Map 2: The road to Tripoli. The solid arrows show the advances that have already been made. The dashed arrow shows the possible route to Tripoli.

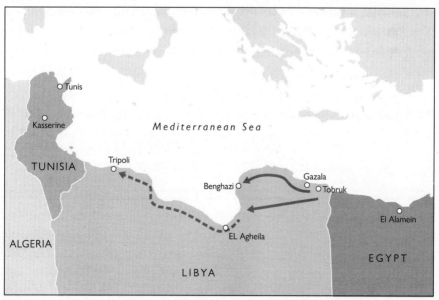

But you would have to act now, before the enemy can pull itself together. There is possibly just enough in the way of captured supplies, fuel and transport to make a quick dart towards Tripoli viable, but it would be a big risk, not least because such a move has only been implicitly, rather than explicitly, sanctioned by Wavell. There is also a newly appointed governor of Cyrenaica, Lieutenant General Henry 'Jumbo' Maitland-Wilson, whose view is not yet known, partly because you do not know where he is.

What advice do you offer O'Connor?

AIDE-MEMOIRE

* There is no clear order to advance on into Tripolitania; you would be acting on your own initiative.

* The enemy seems to be in complete disarray. Delay could let them off the hook.

* It is far from clear that you have the supplies or troops to get to Tripoli. If your forces become isolated, they could be lost and everything you have just won would be put at great risk.

* Capturing Tripoli now would free forces to go to Greece later, or even open up the possibility of an advance into Tunisia.

* You need air and naval support for a dash to Tripoli, neither of which you have secured.

THE DECISION

Should O'Connor act without direct orders, take the gamble and push on towards Tripoli, or should he consolidate the current position and await clarification of the wider strategic position in the Mediterranean?

▶ To push on to Tripoli, go to **Section 2** (p. 80).

OR

▶ To await new orders and guidance, go to **Section 3** (p. 88).

SECTION 2
WEYGAND'S CONVERSION

8 April 1941, Jacques Lemaigre Dubreuil's residence, Algiers, Vichy French North Africa

It is the evening of 8 April in Algiers, the administrative and political centre of Vichy French North Africa, and a group of high-ranking French officials and dignitaries has assembled at the plush residence of Jacques Lemaigre Dubreuil, a well-known French businessman and newspaper magnate. Lemaigre Dubreuil is an ardent right-wing nationalist who has railed against the German occupation of France. When France fell to the Germans the previous year he transferred his base to Casablanca, but he has links and interests across many French territories and has maintained his business links with metropolitan France itself. Tonight he has assembled a group of military and political leaders to meet and assess the developing military situation in North Africa – all of them fired by the collapse of Italian power in Libya at the hands of the British forces commanded by General O'Connor.

In addition to Lemaigre Dubreuil, seated around the table are: Colonel Alphonse Van Hecke, the Flemish commander of the Chantiers de Jeunesse, the French military youth movement; Lieutenant Henri d'Astier de la Vigerie, an army staff officer; and Captain André Beaufre, a veteran of the French colonial wars of the 1920s. You are *Jacques Tarbé de Saint-Hardouin*, the forty-year-old political assistant to General Maxime Weygand, the Vichy government's Delegate-General in French North Africa. You are a well-regarded career diplomat, who was party to the signing of the peace treaty with Germany less than a year ago. Subsequently you were assigned to Algeria as the senior diplomat to aid the Delegate-General, the politically appointed senior figure in French North Africa, who since the autumn of 1940 had been General Maxime Weygand – and it is Weygand whom everyone at the meeting wants to see.

Weygand is a generally mild-mannered man, short in stature, assiduous and unassuming, but you also regard him as astute and considered. You know that he is something of an outsider

3. Jacques Tarbé de Saint-Hardouin

among the French elite, being of uncertain background and parentage; De Gaulle, the leader of the Free French, has even claimed that Weygand is not to be trusted because he was born in Brussels and does not have 'a drop of French blood in his veins'. Yet Weygand will be pivotal to the ensuing discussion and decision, and everyone in the room knows that.

You have travelled to Lemaigre Dubreuil's house for his meeting with Weygand to avoid the prying eyes at Weygand's official residence and headquarters in Algiers, for although Weygand has been loyal to the Vichy regime in France, he is still viewed with a little suspicion by the collaborationist elements back home. It is true that he was an architect and advocate of the armistice with Germany in 1940 and has vigorously prosecuted some of the more extreme measures imposed by the Pétain government, but he was moved to the North African post a few months ago to get him out of the way, as rumours were circulating that Weygand's views on France and its colonies steering well clear of further involvement with Germany's war effort were not going down well in Berlin or Vichy.

But Weygand's determination for Vichy France to remain as independent of Hitler's Third Reich as possible is being tested as never before by the advance of British-led forces into Libya: he is fully briefed on recent events there. On the evening of 7 February 1941 Lieutenant General Richard O'Connor had begun a sudden dash into Tripolitania spearheaded by the armoured cars of the 11th Hussars. It seems as though, without clear contrary orders from Cairo, he had acted on his own initiative and that he decided to continue his advance until told to change course. The risks had been great, but O'Connor was determined to follow up the stunning success of the defeat of Italy's Tenth Army. To O'Connor it had been clear 'that provided the enemy can be prevented from landing reinforcements, there would be little serious opposition to our advance'. Though hampered by logistical problems and initially without clearly secured support from the RAF or the Royal Navy, O'Connor's advance force faced very little opposition, and such as there was quickly melted away. His troops swept into Sirte (some 180 miles into Tripolitania), capturing supplies of fuel, water and equipment. Italian opposition in the area collapsed and the remaining forces in Libya were left in complete disarray. The defence of Tripoli had been entrusted to General Italo Gariboldi's Fifth Army, but although on paper it had four infantry divisions and a newly arriving armoured division to deploy, most of the heavy equipment and artillery had already been transferred to the Tenth Army and lost in the weeks before the Italian collapse in Cyrenaica. There was also limited and only demoralised air support available, and little likelihood of any significant improvement in the Italian position for some weeks to come. The Italian navy, still smarting after being mauled in its home port of Taranto by a Royal Navy carrier-launched air strike the previous year, was most unwilling to risk its remaining major ships without a properly coordinated plan with a good chance of success.

The one potential difficulty confronting O'Connor's forces had been the impending arrival of German units in Tripoli (a possibility to which London had been alerted through the work of the code-breakers at Bletchley Park). But the first German troops only arrived on 12 February and disembarkation had been slowed by Tripoli's limited

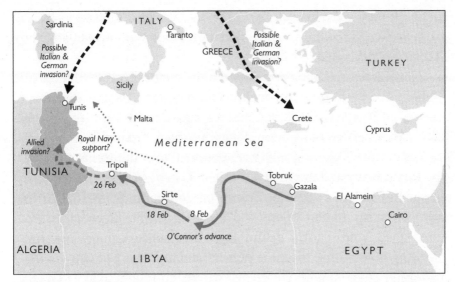

Map 3: Vichy Tunisia, Spring of 1941. The Axis powers have control of Algeria, Sicily and mainland Italy.

ship-handling capacity. The Germans had not sent any tanks at this stage, nor heavy 88mm anti-tank guns. Without any useful such weapons to stop British armour, the Axis forces were at a major disadvantage. The Luftwaffe still had no base of operations in North Africa and therefore had to shuttle aircraft from Sicily to offer any support, seriously limiting its ability to aid the defence of Tripoli.

O'Connor's troops had pressed on from Sirte on 18 February and clashed with the first German units, elements of 5th Light Division deployed to North Africa under the command of the newly arrived General Erwin Rommel. Despite some minor tactical successes, the heavily outgunned Germans were driven back and in the confusion Rommel, commanding too close to the front line, was captured by a British patrol.* Axis opposition soon melted away. Within five days O'Connor's advance units were closing on Tripoli, which descended into turmoil; the remaining senior commanders and political officials began to evacuate, but Royal Navy ships raided the escape routes,

---

* A fate that historically befell O'Connor in April 1941. He spent two years in a prisoner-of-war camp in Italy.

although they were harried by Italian and German aircraft operating from Sicily. As the Italian ships fled north, they were also attacked by Allied aircraft based in Malta.

Nevertheless, the leading Allied units were forced to halt outside Tripoli, due to shortages of fuel and ammunition; the advance was at the end of its logistical tether, and a long trail of broken-down vehicles and exhausted units littered the route back to Sirte. But the panic caused by the lightning advance, coupled with the threat posed by the Royal Navy and the RAF – limited as it was – was enough to see the end of Italian control of Tripoli on 26 February 1941. A further 30,000 mostly Italian troops had been forced to surrender and fall into Allied hands. It was a stunning Allied success, and Mussolini's standing both within the Axis powers and at home had taken a huge battering; coupled with the dismal campaign fought against Greece, political indicators pointed to a possible coup against the Fascist dictator.

In Algeria, prior to the British seizure of Tripoli, you, as Jacques Tarbé de Saint-Hardouin had witnessed Weygand steering a very careful path through the murky waters of Vichy–German–British diplomatic relations. Churchill had been trying to woo Weygand since the autumn of 1940, and although the British had never been rejected outright, Weygand stuck to the neutral Vichy policy. Like you, Weygand was well aware that French forces in North Africa were painfully ill-equipped and underprepared for any conflict and that the British were in no way to be trusted – their military failings were self-evident. Churchill had despaired of Weygand and by February 1941 was on the point of cutting off any real lines of communication with him.

Yet a few weeks later, all had changed. You were with Weygand when news of O'Connor's great advance filtered through to Algiers, but you were also present when rumblings of uproar in Vichy emerged soon afterwards. It seems as though Berlin is pressing for a new set of agreements – called the Paris Protocols – to be agreed between Vichy and Germany. You have heard that Hitler wants Pétain and the Vichy regime to allow German and Italian troops and equipment to be sent into Vichy-controlled Tunisia to hold back any further British advance or even to prepare a counter-strike to retake Tripoli. When you met

Weygand to discuss this development, he had been apoplectic: it totally contravened the avowedly neutral and semi-independent stance he insisted had been at the heart of the armistice with Berlin the previous year. To allow Italian and German troops and equipment to be based in Tunisia simply drew French colonial territories into the war. But what could be done to stop it?

It was then that you received a covert message from Jacques Lemaigre Dubreuil that he and other leading members of the French ruling group in North Africa wanted to discuss the developing situation. As you sit and listen to them, it is clear that the British have been in contact – they are preparing a force of mechanised ground troops, supported by air and naval assets, to race into Tunisia if Weygand declares for the Allies and blocks any German or Italian incursion into North Africa. You watch Weygand's reaction – he does not reject the idea out of hand, something he would have done just a few weeks earlier, but he is still cautious. What size of force would be committed? Did the British know how parlous the state of Vichy military strength was in North Africa? If the Germans descended in force, how quickly could the British get to Tunis? Did the British know that Weygand could probably not count on bringing everyone with him? There would be opposition, including from Admiral Jean-Pierre Esteva, a confirmed supporter of the Vichy government, and the governor of Tunisia.

Weygand announces that he needs more time to consider the strategic implications of joining the Allies. He also requires more information on the size of British commitment, though you note that he is at least glad the so-called Free French forces led by De Gaulle have not been much mentioned – he has little time for them.

As you drive away, what do you advise?

AIDE-MEMOIRE

```
* The British, despite their recent success
  against the Italians, have hardly showered
  themselves in military glory since the war
```

began. Can they be trusted with this dynamic plan?

* The Germans have to be prevented from landing in Tunisia - it would be a political disaster for France if they moved in.

* The rumours of the Paris Protocols are at this stage just that - rumours.

* The British-headed coalition offers a much more palatable long-term ally for France, but is the time now right to join them?

* Acting prematurely could see all of France directly occupied by the Germans and there would be no chance of resisting it. If Weygand acts first, by siding with the British, there is no doubt that the rest of France will pay dearly.

* The forces available to Weygand are slight, and if the Germans and Italians arrive in North Africa in strength there would be little Weygand could do to stop it, unless the British send sufficient troops to help. But will they?

THE DECISION

Should Weygand agree to the British move into Tunisia and support it, or should he stay neutral and hope that the Germans do not invade Algeria and Tunisia themselves?

▶ To stay neutral, conserve what you have and wait to see what emerges, go to **Section 4** (p. 96).

OR

▶ For close cooperation with the British, go to **Section 5** (p. 100).

SECTION 3
GREEK TRAGEDY – THE ROAD TO EL ALAMEIN

5 August 1942, Ruweisat Ridge, El Alamein, Egypt

Winston Churchill, the British Prime Minister, has just arrived at the field headquarters of the Eighth Army, sited a little behind the Ruweisat Ridge south of El Alamein. He is here to confer with General Claude Auchinleck, the commander of the Middle East, who has recently been forced to take direct control of the British Army fighting the Axis forces in the deserts of North Africa. The Prime Minister is in a bad mood, rattled by the gloomy and dejected atmosphere in Egypt that he has thus far encountered since beginning his tour. You have accompanied the Prime Minister to El Alamein as you are *General Alan Brooke*, the fearsome Chief of the Imperial General Staff (CIGS), in effect the head of the entire British Army. Since the spring of 1942 you have been the chair of the Chiefs of Staff committee and thus have become Churchill's closest senior military advisor. You are recognised and feared for your forceful personality and commanding intellect, which hardly matches your fascination with the much gentler pastime of ornithology. You took over as CIGS in December 1941 in some of the darkest hours of the war.

The two of you are in the Middle East to appraise the parlous situation and make any necessary adjustments that you see fit, although ultimately it will be Winston who will make the final decisions, though he relies heavily on your guidance. You have a tempestuous relationship with Churchill, a man you admire greatly for his many qualities of inspirational leadership, energy and determination. Yet, equally, you recognise his mercurial nature, his tendency to act impulsively and his willingness to act without always considering the consequences. You believe that he admires you for your ability to confront him with bullish determination; he seems to recognise that you will tell him what you think, rather than what he wants to hear. Your partnership has gone well enough since you took up the job as CIGS, but it is being tested by the gravity of the situation you are confronting in the Middle East, and in particular in Egypt.

4. General Alan Brooke

Some eighteen months have passed since Churchill and the War Cabinet Defence Committee had ordered O'Connor to halt his advance deeper into Libya. Just before that fateful decision was made, General Archibald Wavell had signalled at around 1900 hours on 10 February 1941, to the then CIGS, General Jack Dill, back in London, his desire to press on to Tripoli as soon as possible. Wavell knew that O'Connor was ready to advance, though as yet he did not know his exact whereabouts. Wavell signalled: 'Extent of Italian defeat at Benghazi makes it seem possible that Tripoli might yield to a small force if despatched without undue delay.' But, to Wavell's dismay, Dill had replied the following day with the frustrating news that Churchill's War Cabinet Defence Committee did not agree. Dill tried to raise the matter once more, but was met with a furious outburst from Churchill. 'I could see the blood coming up his great neck and his eyes began to flash,' Dill had recorded later. Churchill had lambasted Dill, claiming that Wavell had 300,000 troops in his command and that he would have to find the required resources to aid the Greeks. The Defence Committee then sent very clear instructions to Wavell, stating unambiguously that their decision ruled 'out any serious effort against Tripoli ... you should ... concentrate all available forces in the [Nile] Delta in preparation for movement to Europe [Greece].'

This strategic choice had not played out very well, however, and Britain's entire strategy in the Mediterranean had quickly unravelled. Churchill had been determined to keep the Axis powers out of the Balkans since the outbreak of the war, and had planned to establish a pro-Allied Balkans front to thwart Hitler. The Balkan powers had wanted to keep out of any conflict, but by late 1940 events were starting to sweep such intentions aside. The Italians attacked Greece from Albania (which they had occupied in 1939) and Germany drew Hungary, Romania and eventually Bulgaria into the Axis powers. Yugoslavia, prompted by a British-backed coup, withdrew from such entanglements with Berlin, but this simply brought down Hitler's fury.

The Greeks had driven Mussolini's initial invasion back, but when German troops started arriving in Romania, and Bulgaria seemed likely to allow a similar use of its territory, the Greeks asked the British for serious help. Wavell, under orders from London, despatched a strong force of some 60,000 troops to support the Greeks, but when the Germans swept through Yugoslavia and then invaded Greece there was little the British and Greeks could do to stop them. The Germans combined their skills in manoeuvre and speed to seize the initiative, while the British and Greeks failed to coordinate their strategies effectively. Three weeks later German troops were in Athens. Wilson was forced to begin an evacuation of the mainland, but the Germans harried the Allied forces as they withdrew. Although the Allies managed to withdraw 50,000 troops, twenty-six troop-laden ships were lost to enemy action, and close on 8,000 soldiers were lost to captivity. By the end of May the important island of Crete had also fallen to the Germans, threatening British interests in the Eastern Mediterranean.

The entire Greek campaign had been a costly, humiliating disaster and a major strategic blunder. Worse was to follow for Wavell, however. His control of Eastern Libya was soon totally unhinged by a dynamic German attack commanded by the newly arrived German commander Erwin Rommel, an attack that had begun merely as a 'reconnaissance in force'. The Allied forces in Libya, stripped of many units and significant air support, were driven back in disarray, surrendering their prizes won from the Italians just a few months earlier. Wavell was eventually replaced by Lieutenant General Claude Auchinleck, who oversaw a major reorganisation that saw the creation of the Eighth Army, which then launched a major offensive against Rommel's forces in the autumn of 1941. Alas, despite some initial successes, the Allied position again deteriorated and by the summer of 1942 the Allies had been driven back further than ever before into Egypt.

In June, Auchinleck had assumed direct command of the Eighth Army (he had been forced to sack the first two generals he had appointed) and had authorised a further withdrawal to defensive positions at El Alamein, only some sixty miles from Alexandria and the Suez Canal. Britain's entire grip on the Middle East hung in the balance, and Cairo was in turmoil – a panic that became known as 'the flap'. Preparations began for the possible evacuation of Cairo, but Auchinleck's troops fought desperately and held off Rommel in July at the First Battle of El Alamein. The British-led forces even launched a series of counter-attacks, but they were poorly coordinated and had came to nothing, save cementing the view that Auchinleck's command was out of ideas and low on morale. It was also clear that the Allies were not out of the woods, either, and that Rommel's army – despite

Map 4: The German advance to El Alamein, Spring-Summer 1942

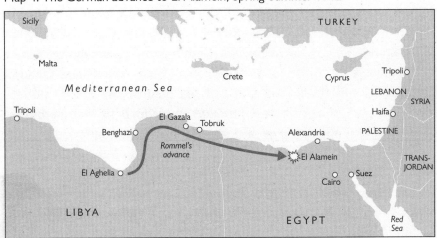

being outnumbered and operating at the end of a very tenuous supply line stretching back across Libya – would attack again soon.

You had seen Churchill's confidence in Auchinleck drain away by the summer of 1942, despite your best efforts to talk Auchinleck up. Winston had insisted on following you to the Middle East to evaluate the situation, and you had travelled together to meet Auchinleck at his advance headquarters to get a sense of the situation. On arriving, Churchill's grumpy mood is quickly noted by everyone present, a situation not improved by swarms of flies and an unappetising breakfast of fried food. You are hopeful that you can save Auchinleck or allow a balanced judgement, but although you have defended him to Churchill, you are no longer so sure. You write in your diary: 'It is very trying having to fight battles to defend the Auk, when I am myself doubtful why on earth he does not act differently.'

Once at Eighth Army HQ south of El Alamein, a disastrous briefing from Auchinleck confirms your own and the Prime Minister's worst fears. You meet in the Auk's operations caravan, Churchill thrusting 'stubby fingers' at the map, asking why units could not act immediately. He wants prompt, dynamic action, but Auchinleck and his chief staff officer, the mercurial Lieutenant General Eric Dorman-Smith, spend the entire time telling Churchill why things cannot be done. It is not the right way to manage the Prime Minister at all, and it seals their fates.

You agree with Winston that Auchinleck has to go, but who will replace him? One issue is straightforward: the easy-mannered and diplomatically astute General Harold Alexander is best suited to replacing Auchinleck as overall commander of the Middle East in Cairo, but who should you appoint as the new Eighth Army commander?

There are two distinct possibilities. One is Lieutenant General Bernard Montgomery. You know of his qualities as a keen professional with a strong grip on command and leadership – something it seems the Eighth Army is lacking. Monty is currently only commander of the South-Eastern Army defending Kent, Sussex and Surrey, but he has built a reputation for being outspoken, determined and single-minded, as well as being abrasive and rude. Eighth Army lacks direction and purpose, and requires firm and authoritative leadership to drive it forward; Monty is surely a good fit for the post. In fact Auchinleck, despite having had a number of run-ins with Montgomery in 1941, has already agreed with you that Monty would be a suitable appointment to command the Eighth Army, despite not having had any experience of desert warfare. Appointing Monty would be a risk, but he would bring in a new approach and fresh thinking.

The second option – and the one initially favoured by Churchill – is Lieutenant General William 'Strafer' Gott, currently commander of XIII Corps positioned at El Alamein. Gott is a big man with a big personality, well liked by his troops and with tremendous experience of desert warfare, having been in the Middle East since the late 1930s and in combat with the Axis forces since the outbreak of the North African campaign. He is brave, aggressive and outgoing, exactly what might be needed for the Eighth Army. The Prime Minister met Gott shortly after your meeting with Auchinleck and has formed a good impression of him – a sharp contrast with the pessimistic and negative outlook of the Auk. 'I convinced myself of his high ability and charming personality.' But Gott has been fighting in the desert for years; he might be popular and impressive with lots of relevant experience, but does he have the ability to seize the moment?

As you arrive in Cairo to discuss the matter, what should you suggest to the Prime Minister?

AIDE-MEMOIRE

* Montgomery brings fresh ideas and fierce professionalism.

* Monty has no experience of desert warfare and is known to be abrasive and awkward. Would his personality suit the emotionally drained commanders and soldiers of the Eighth Army?

* Gott has a wealth of experience and is popular with those he commands and those he works with. Is this not a time to pull everyone together and rebuild morale?

* Gott has a big and powerful reputation and good standing - surely what the Eighth Army now needs?

* Monty brings single-minded determination, with a careful eye for meticulous planning. Don't you need someone with new ideas and plenty of confidence in them?

* Monty is known for his caution, but Gott seems determined to strike soon against Rommel and wrest the initiative from the enemy.

* Might Gott be tired, however? Has he not tried out most of his ideas on the enemy already, without obvious success?

THE DECISION

Who should take over as commander of the Eighth Army in its perilous position? Rommel is soon to launch a further effort at breaking through to Alexandria and the Suez Canal. The Eighth Army has more troops and equipment, but such superiority has not helped before, when Rommel dominated the battlefield with his drive and determination. It is essential that Churchill and you pick the correct commander – another false choice could prove calamitous. Who should you choose: Monty or Gott?

▶ If you recommend Montgomery, go to **Section 6** (p. 106).

OR

▶ If you recommend Gott, go to **Section 7** (p. 112).

SECTION 4
PROPPING UP THE BALKANS

16 May 1942, Ankara, Turkey

The government in Turkey is in a state of high tension. Diplomats bustle about, conveying messages from the British, the Americans, the Soviets and the Germans; Turkish foreign policy is being buffeted by conflicting pressures and the next few days could decide the future path of Turkey's place in the world, and quite possibly its independence and security. You are the Turkish President, the fifty-six-year-old *İsmet İnönü*.

You are an educated technocratic leader who speaks fluent Arabic, English, French and German as well as Turkish. You are a military man and had previously been intimately involved in the wars and conflicts that engulfed Turkey in the post-First World War years, leading to the creation of the republic under the leadership of the charismatic Mustafa Kemel Atatürk. Since 1938 you have been President and you have little desire to see Turkey – still only just putting itself back together after the collapse of the Ottoman Empire – embroiled in any war. Yet Turkey is fearful of Italian expansion into the Balkans from the west and Soviet threats from the east, particularly Stalin's desire to place Soviet troops and bases in the Dardanelles. When

5. President İsmet İnönü

the war broke out, your government had initially sided with the Allies, but despite a treaty that should have resulted in Turkey acting militarily when Greece was attacked by the Axis powers, you have kept Turkey resolutely neutral.

But the situation is shifting. By the beginning of March 1941 Libya had fallen to the Allies and North Africa had been freed of Axis fighting forces. The British had focused their efforts on capturing Tripoli in February and had only very recently shown much interest in supporting the Greeks against the Italians and Germans – it was all too little, too late. There had been few Commonwealth forces in mainland Greece when the German invasion came on 12 March, and Greek opposition, dismayed at the lack of support from the wider world, crumbled after a short campaign. An attempt by the Germans to use airborne forces to spearhead an invasion of Crete at the end of April failed dismally, however, as the weight of Allied forces funnelled onto the island by the British had fended off the paratroopers; losses were heavy on both sides, but when the fighting on Crete ended, many thousands of German airborne troops were captured.

But Britain's lacklustre support to the Greeks has shaken your faith in their commitment and ability to contain the Axis powers. Could, or would, they be any more supportive if the Axis powers – or indeed the much-feared Soviet Union – started applying pressure on Turkey? Could Churchill be trusted in any way? Late in February 1941 the British Foreign Secretary Anthony Eden and the Chief of the Imperial General Staff, General Jack Dill, had visited Ankara in an effort to bolster Britain's standing and to encourage you to act more aggressively to defend Turkey's interests. It was clear to you and your staff, however, that Britain had taken its eye off the ball in the Balkans; there was little faith in British military or political commitment.

As British standing in the Balkans has faded, Germany has emerged as the power broker in the region, and the German ambassador to Turkey, Franz von Papen (who had been instrumental in elevating Hitler to power in 1933), has been working to bring Turkey into Germany's orbit. You are aware of a growing mood from some political factions in Ankara that a more pro-German stance should be adopted. It seems that some in Berlin even believe that Turkey might join the Axis, but you have convinced von Papen that is highly unlikely. Nonetheless, you have signed a non-aggression pact with the Germans to keep them happy, even though you know it will annoy London; and indeed the Americans, who are offering supportive words, though as they are neutral, there is little practical help they can provide.

Now, however, matters have come to a head. A few days ago news broke of an anti-British rebellion in Iraq and the leaders of the coup are requesting military aid from Germany. Hitler is willing to help, and neutral Vichy French territory in Syria is being used as a staging point to send some aircraft to Iraq. Franz von Papen has now arrived at your presidential residence to discuss the possibility of Turkey allowing its territory to be used to transport German ground forces and other supporting units to Iraq, to offer much greater levels of support to the rebels. Clearly this would rile London no end, but would also grow ever more positive links with Berlin.

For a few days Berlin bombards you with requests, sometimes bordering on demands, while London and Washington urge you to refuse. Moscow is also sending veiled threats. Anthony Eden himself travels to Ankara to persuade you to stay completely neutral, but he also tells you in great secrecy, and off the record, that the British are convinced the Germans are about to attack the USSR. Do you want to risk getting caught up in what will obviously be a titanic and appallingly destructive confrontation? Helping the Germans now could draw you into a war with the communists of Stalin, surely a disastrous scenario?

You and your government are not minded to allow the Germans what they want, but the prospect of getting caught between Berlin and Moscow fills you with dread. You stall von Papen's requests and, a few days afterwards in late May, the Germans attack the Soviet Union. You are quite happy to see your two greatest security worries battling it out and you stay resolutely neutral to await events. Your government refuses Germany's request for transit rights and the rebellion in Iraq is soon snuffed out.

AFTERMATH

Although the Russo-German war in 1941 was a close-run thing, the Soviet Union survived, and the two belligerents became locked in a titanic four-year struggle, in which President İnönü was resolutely opposed to becoming embroiled. Turkey settled back into its balanced, carefully navigated neutral route until the outcome of the war was assured in 1945, which brought an opportunistic declaration of war on Germany. Turkey suffered economically during the war years, but when heckled in later life that he had allowed his people to go without food, İnönü replied curtly, 'Yes I let you go without food, but I did not let you become fatherless.'

THE DECISION

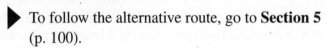 To follow the alternative route, go to **Section 5** (p. 100).

OR

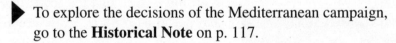 To explore the decisions of the Mediterranean campaign, go to the **Historical Note** on p. 117.

SECTION 5
THE BATTLE FOR TUNISIA

28 April 1941, bridge of HMS *Warspite*, Central Mediterranean

Senior naval officers and personnel are readying for action on board HMS *Warspite*, the flagship of the Royal Navy's Mediterranean Fleet. Some distance to the north-west lies the Italian navy, which is escorting a convoy of enemy transport and supply ships headed for Tunis. You are **Admiral Andrew Cunningham**, the highly regarded fifty-eight-year-old Commander-in-Chief of the British Mediterranean Fleet operating out of Alexandria. With more than forty years of experience in the navy, and a string of battle honours stretching from the Dardanelles in the First World War through to the aircraft-carrier attack on the Italian fleet at Taranto in November 1940, you are held in great esteem. Now another major task has been assigned to your fleet: the interception of enemy transport ships carrying reinforcements and equipment to Tunisia. Although the Italian navy is out in strength and you are operating within range of enemy land-based aircraft, it is essential that the Germans and Italians are

6. Admiral Andrew Cunningham

7. HMS *Warspite*

stopped from moving forces into North Africa, a region in turmoil following the attempted overthrow of Vichy control.

You are fully briefed on the unfolding situation in Tunisia, where Allied control hangs in the balance. Just a few days earlier you had been alerted that General Maxime Weygand, the Vichy government's Delegate-General in North Africa, was about to declare for the Allies and against the collaborationist Vichy regime. His decision had been precipitated by the arrival of Allied troops (led by Richard O'Connor) in Western Libya, and the Italo-German initiative to send troops into neighbouring Tunisia to hold back or possibly even attack the British. This had been too much for Weygand and he signalled his intention to join the Allies and bring French forces in Algeria and Tunisia with him. For the coup to succeed, it would be vital that the British push forces quickly into Tunisia, as time would be of the essence before the enemy reacted. In addition, Weygand could not guarantee that all his forces, which were in any case small in number and limited in capability, would switch with him. Some would still maintain their allegiance to the French government in Vichy; the pro-Vichy Admiral Jean-Pierre Esteva, the military commander in Tunis, was especially not to be trusted. It was also highly likely that the Germans and Italians would invade Tunisia in response to Weygand's declaration; so speed would be of the essence.

Churchill had been much animated by the secretive and clandestine nature of the plan to wrest control of Tunisia and Algeria from Vichy and he enthusiastically backed the operation. O'Connor readied fast-moving troops on the Tunisian–Libyan border, the RAF flew in extra aircraft from Egypt, and the Royal Navy's Mediterranean Fleet, under your command, had sailed out of Alexandria bound for the dangerous waters of the Central Mediterranean. Blocking the movement of enemy forces from Europe into North Africa was your primary responsibility.

On 17 April Weygand had issued orders to seize all port and airfield facilities in Tunisia and Algeria to deny them to the Axis forces, should they descend. Pétain immediately countered by sacking Weygand, ordering his immediate return to France; the pro-Vichy Admiral Esteva, based in Tunis, was ordered to take charge. This immediately set in motion a struggle for control of Tunis, with Weygand's forces holding some key positions, while Vichy loyalists held others. Into this unstable situation a small German airborne force had arrived over Tunis with orders to link up with Esteva's forces and seize control of airfields and the port. With the Luftwaffe and the *Regia Aeronautica* (the Italian air force) controlling the skies, there was little the French forces could do to resist, but many switched to Weygand when they saw German troops arriving once more on French-controlled territory. Fierce battles broke out in Tunis and Sousse.

As events were unfolding in Tunisia, O'Connor's troops began their advance across the border from Libya. The total Allied force was quite small because the capacity of Tripoli to supply an army was limited by the huge distances involved in linking it to the main British bases back in Egypt. O'Connor's forces were also still spread thinly across North Africa, despite the reinforcements provided by Churchill.

The Vichy French border troops, under Weygand's direct orders, allowed the Allies through the border and O'Connor's forces were soon racing north to seize the Mareth Line, a system of defensive positions at a geographical bottleneck in south-eastern Tunisia, between the Matmata Hills and the Mediterranean Sea. Although the defences had been partially dismantled since the establishment of Vichy, they could still present a problem if opposing forces held them, but O'Connor's troops swept through without resistance.* Some French units were now openly switching to the Allies, and Weygand's troops still retained some control of Tunis and Sousse.

Vichy French troops in Sfax, however, stuck to their orders from Admiral Esteva and refused to surrender the town to the Allies. O'Connor was presented with the decision as to whether to fire upon French troops. Meanwhile the Italians and Germans were flying in a steady trickle of troops and equipment, and on 24 April Weygand reported that the enemy had secured the port facilities in Tunis. The enemy would undoubtedly now start trying to ship in reinforcements.

Enigma intercepts alerted the British to the German-Italian plan to send a convoy of five troop ships complete with stores and equipment into Tunis. They had to be stopped. Your orders quickly arrived instructing you to intervene. It was highly likely, you were advised, that the Italian fleet would also come out to support the troop ships. It was estimated they would be able to muster a battleship, six to eight cruisers and close on twenty destroyers, all probably under the command of Admiral Angelo Iachino. They would also be supported by local air forces.

You command three battleships, an aircraft carrier, seven cruisers and seventeen destroyers, and you also know that Force H, a fleet operating out of Gibraltar, will be acting in support from the west. Will

---

* This was the scene of a battle in North Africa in 1943, won by Monty's Eighth Army before it pushed on to Tunis itself.

it be enough against the enemy fleet and the enemy air forces operating out of Italy?

Martin Maryland reconnaissance aircraft from Malta have now spotted the Italian transport ships and you quickly order an air strike of Fairey Swordfish torpedo-bombers from HMS *Illustrious*. You come under enemy air attack, but damage is slight and your air strike then reports that they have left two transport ships ablaze. Shortly afterwards reconnaissance aircraft report that the Italian fleet is steaming to intercept, and air attacks against your force intensify – two destroyers and a cruiser are damaged and are forced to withdraw. You then contact the Italian fleet, and a running battle breaks out in which the heavy guns of your three battleships prevail, driving the Italian fleet back with some considerable losses.* Enemy air attacks have started to become a major concern and when one of your battleships, HMS *Barham*, suffers some serious bomb damage you are forced to withdraw. However, it later transpires that three enemy destroyers and a cruiser have been sunk, while the Italian battleship *Vittorio Veneto* has been put out of action for many months.

AFTERMATH

A further air strike from HMS *Ark Royal*, part of Force H, sank another enemy transport ship and left yet another ablaze. Total damage to the Axis seaborne expedition had been calamitous. Only one of the transport ships arrived in Tunis unscathed, and many troops and much equipment had surely been lost. Hitler raged at the incompetence of the Italians, whom he naturally blamed for the failure of the mission.

But Axis control of the skies over Tunis was turning the battle in their favour. More troops were flown in, and Weygand was eventually forced to flee the capital with some of his troops. He went on to link up with O'Connor, who had advanced to Sfax. By May the Allies held Sousse, and the Axis forces Tunis. Both sides lacked the resources to

---

* This was a similar engagement and outcome to the Battle of Cape Matapan, fought historically in March 1941.

press the issue further and Hitler's attention soon became focused on Operation Barbarossa, the invasion of the Soviet Union. For the British, their gaze was drawn to the Eastern Mediterranean, where Greece had collapsed and Turkey appeared to be teetering on the brink of joining the Axis.

THE DECISION

▶ To follow the alternative route, go back to **Section 4** (p. 96).

OR

▶ To explore the decisions of the Mediterranean campaign, go to the **Historical Note** on p. 117.

SECTION 6
MONTY'S ALAMEIN

Midnight 30 August 1942, Montgomery's Eighth Army tactical HQ, Burg el Arab, Mediterranean coast of Egypt

Over the last few hours German and Italian forces have been intensifying their attacks against Allied Eighth Army positions in the desert to the south of El Alamein. It is an offensive the British have been expecting for many weeks, an offensive that could well determine the outcome of the entire North African campaign. If the Allies don't hold back Rommel's forces and Axis troops break through, they could be in Alexandria and Cairo within a few days. The new Commander-in-Chief of the Eighth Army, Lieutenant General Bernard Montgomery, has been given the unenviable task of stopping Rommel and then driving him back. You are ***Brigadier Francis 'Freddie' De Guingand***, Montgomery's Chief of Staff. You have known Monty for many years and have a clear insight into what makes your new commander tick. Your colleagues among the senior Eighth Army staff have little knowledge of Montgomery – he is a rather obscure figure to many of them. In complete contrast, however, you are

8. Freddie de Guingand

well known and popular, someone who communicates well and engenders camaraderie. You have recently taken over as Chief of Staff in the Eighth Army after a stint leading the intelligence staff in Cairo, but when Monty arrived to take over command of the Eighth Army in mid-August, you feared he would sweep out many of the existing senior staff – you included – to freshen things up. Yet he has affirmed his faith in you.

Now you have to inform him that the expected German offensive has indeed begun. Monty has retired to his caravan to sleep and when you wake him to tell him the important news, he simply murmurs, 'Excellent, excellent' and goes back to sleep. You have been getting reacquainted with Monty's idiosyncrasies after some years of following different army career paths, but this will no doubt make a great anecdote in years to come.

Ironically, Montgomery had only become commander of the Eighth Army in tragic circumstances. A few weeks earlier, and despite the efforts of General Alan Brooke, Chief of the Imperial General Staff, to dissuade him, Winston Churchill had selected Lieutenant General William 'Strafer' Gott to take over as commander of the beleaguered Eighth Army and had signalled his decision to the War Cabinet in London. Consequently, at 1610 hours on 7 August 1942, at Burg el Arab airfield in the deserts of Egypt, Gott had boarded a Bristol Bombay transport aircraft bound for Cairo. The aircraft had been packed with the wounded, press correspondents and other officers. Alas, tragedy struck, for shortly after take-off the Bombay was set upon by two German Messerschmitt Bf 109 fighters and forced to crash-land. Though one of the Allied pilots and three of the passengers managed to escape, fifteen were caught inside the aircraft when it blew up and were killed; William Gott was not one of the survivors. The Bombay had been flying on the same route that Churchill had taken a few days earlier – a route considered safe. It was a fluke that the German fighters had been driven into the air space taken by the Bombay, but the shooting down of the aircraft set in motion events that would elevate Bernard Montgomery to the position of the most celebrated and well-known British military commander of the Second World War.

Monty had always been Brooke's choice to take over command of the Eighth Army and, with Gott now gone, it fell to Montgomery to assume command. He had swept into Cairo on 12 August, determined to stamp his authority on his new army. He met Auchinleck to get up to speed, then Alexander (his new theatre boss), before meeting you: 'You chaps seem to have been making a bit of a mess of things,' he quipped.

Monty has brought grip and clarity to the command of the Eighth Army since his arrival. He has not radically altered the plans that had been formulated by his predecessors for the defence of the El Alamein positions (particularly the vital Alam el Halfa ridgeline), but he has invigorated the staff and fostered a new positive attitude. When he first arrived he saw all the senior staff for a pep talk. At first glance, he did not cut an impressive figure – slight, with white knees and a high-pitched dogmatic voice – but his words caught everyone's attention. He held up a piece of paper: 'A charter from the Prime Minister. Hit Rommel for six out of Africa . . . Here we will stand and fight; if we can't stay here alive, then let us stay here dead.' There would be no withdrawals, defeatist talk would cease and there would be no more 'bellyaching' about things that could not be done.* He then assumed command of the Eighth Army and set about reorganising the command structure, links with the RAF and the headquarters organisation. You regard Monty's arrival as an exhilarating breath of fresh air and a real example of the 'Projection of Personality', something in which Monty is clearly a great believer.

Monty's approach is underpinned by good Ultra intelligence derived from the code-breakers at Bletchley Park, a realisation that resource superiority is on your side as his army is growing stronger, and that much of the planning groundwork has already been carried out by Auchinleck and Gott. In fact Montgomery has inherited a defensive plan that has effectively identified the route that Rommel's attacking tanks will take, should the Germans and Italians attack again. Contentiously, Montgomery claims that he still spotted weaknesses in the plan and has made critical changes, though these seem

---

* When Montgomery later claimed in his memoirs that Auchinleck had defeatist plans for withdrawal that Monty simply binned to instil confidence, Auchinleck threatened to sue. Monty had to insert a clarification to avoid litigation.

slight to you. He has, however, strongly emphasised adopting a defensive posture, has withheld transport that might be used to withdraw too easily and has issued orders that Allied tanks must stick to a tactical role of acting as anti-tank guns rather than trying to manoeuvre.

Map 5: The Battle of Alam El Halfa Ridge, August-September 1942

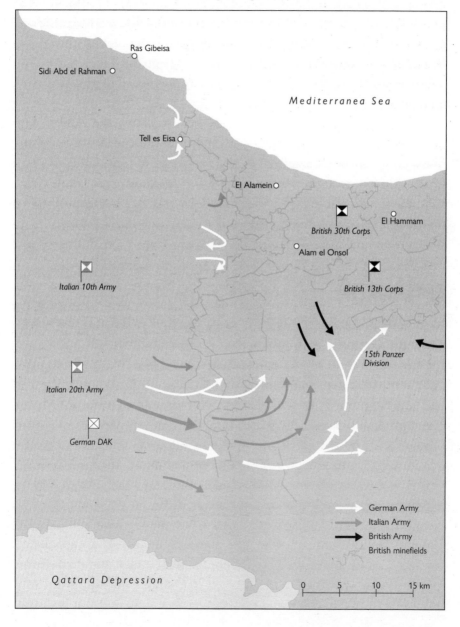

From 30 August until 5 September Rommel's forces probed and prodded the Allied positions in a series of attacks, but Montgomery would not be drawn and used his defensive lines and air forces to rebuff the enemy while inflicting damaging losses on them. After a week Rommel was back where he had started and had suffered the loss of fifty tanks and 3,000 troops. Monty lost more tanks, though proportionately not as many as the Allies usually did, but the positions held and Rommel's sting had been drawn. Later it transpired that Rommel, so used to inflicting crippling losses on the Allies when they overreached themselves, bemoaned Monty's caution: 'The swine isn't attacking,' he complained.

The Battle of Alam el Halfa had been won by the Allies. The defensive line remained intact and although there was some grumbling about Monty's caution in not following up the battered Axis forces quickly enough, when it might have delivered even greater success, the position in Egypt was secured. Now it was a matter of going on the offensive and driving the Axis forces back into Libya.

AFTERMATH

Montgomery remorselessly built up his forces over the next few weeks while the Germans and Italians dug in, sowed minefields and hoped for the best. While Monty enjoyed the arrival of new Sherman tanks, better anti-tank guns and Spitfire fighter aircraft, Rommel became ill and was temporarily invalided home. Not until 23 October did Monty feel it prudent to attack, by which time the Germans and Italians were outnumbered by almost two-to-one in nearly every category of weapons. Starting with a huge 1,000-gun bombardment, the Eighth Army slowly and remorselessly whittled away the Axis forces, but it was by no means plain sailing and, after a day, little progress appeared to have been made. Matters were aided when the German commander, General Georg Stumme, who was standing in for the incapacitated Rommel, came under fire and suffered a heart attack and died. Rommel eventually returned to the fray on 26 October, but was ultimately unable to stem the losses and on 2 November signalled that his army was spent. Although Hitler attempted to force him to stand, within three days the German and Italian forces were in full retreat across Egypt, through Cyrenaica and back into Tripolitania (see Map 2 on p. 77).

Monty was unwilling to commit to an all-out pursuit and preferred an orderly advance, picking off the enemy when the opportunities arose; he was determined not to let Rommel off the hook by overcommitting and risking his advantage. Critics later claimed that a more determined pursuit might have completed the destruction of the Axis forces sooner, but in early December Monty's army broke through the Axis forces at El Agheila and pushed on to Tripoli by January 1943, almost two years after Richard O'Connor had fleetingly considered doing the same.

THE DECISION

▶ To follow the alternative route, go to **Section 7** (p. 112).

OR

▶ To explore the decisions of the Mediterranean campaign, go to the **Historical Note** on p. 117.

SECTION 7
GOTT'S ALAMEIN

31 August 1942, Eighth Army tactical HQ, Ruweisat Ridge, south of El Alamein

It is early afternoon at the height of a battle that could well determine the future of Allied control of North Africa and the Middle East. The tactical headquarters of the Allied Eighth Army, deep in the deserts of Egypt close to El Alamein, is a hive of activity as news and reports of the German and Italian offensives against the Allied positions to the south come flooding in. The timing of Rommel's attack, and indeed its likely route, has been predicted by the senior staff of the Eighth Army and careful plans have been put in place to defeat it. Now, after a tumultuous few hours, it seems as though the Axis forces are running out of steam.

Lieutenant General William 'Strafer' Gott, the newly promoted Commander-in-Chief of the Eighth Army, is particularly bullish about the situation and is carefully ruminating over his forces' next move.

You are *Lieutenant Colonel Charles Richardson*, the thirty-four-year-old chief planning officer at the Eighth Army, responsible for the formulation of future plans and options for the Allied forces. Although you are a fairly recent

9. Charles Richardson

appointment, you have been involved in the final stages of putting together the pieces of the plan to stop Rommel and now it all seems to be going well. Gott took over as Eighth Army commander in early August following the dismissal of Auchinleck. You heard that Churchill had been very impressed with Gott, who was indeed a strong and convincing personality, much liked by those around him – he was a 'soldiers' soldier' and he had been with them through thick and thin in the desert. Most importantly, he also appeared to have a firm grip on the defence plan to repel Rommel's attack. But did he bring anything new to the leadership of the army? How and why would he be able to stop Rommel now, when he and his other senior commanders had failed so often in the last few months? There were whispers about exhaustion, and some considered Gott to be tired and out of ideas, while others thought him a decent middle-ranking commander but out of his depth at a senior post. You are aware of all of this.

Gott's forces had more tanks and guns, greater strength in the air and more troops available, but what they lacked was real self-belief. They had been driven back remorselessly since the spring of 1942, further back than at any point in the war so far – defeat following defeat. Senior and middle commanders despaired at the confusion and indecision of the army command and leadership. They liked Gott, but he needed to step up and meet the challenge of the moment. Would he make the correct choices at the crucial moments? By the early afternoon of 31 August one such moment appeared to be looming. You wondered whether your doubts would be dispelled.

On 30 August 1942 Rommel had launched his breakthrough offensive against the Allied forces at El Alamein. Time and resources were surely against him, but Rommel always seemed to pull

10. William 'Strafer' Gott

something out of the bag to confound the British. Like others around you, you were well aware that Gott had organised a defence based on holding the high ground of the Alam el Halfa Ridge to the south of El Alamein, which was considered the only likely route Rommel's attack could take. Rommel's attack had begun well, spearheaded by the 15th and 21st Panzer Divisions, while the RAF had played only a limited role in trying to hold them back. By the morning of 31 August concern was growing at Eighth Army HQ forces, but in the ensuing hours a 'to and fro' battle had developed. Rommel's attack appeared to be faltering and his leading elements began to fall back.

Gott, a determined and aggressive commander, saw an opportunity to deliver a decisive blow to the Axis forces and issued orders for Allied armoured forces to advance and attack the retiring Germans. You wondered whether this was wise, but moments like this had been few and far between when fighting Rommel, so surely, some of your colleagues argued, this was the time to strike?

Initially there was some success and the German and Italian response appeared confused, but the Allied tank losses to German gunfire soon began to mount. A confused melee continued throughout the night, but the following morning a thrust by the 15th Panzers into the eastern flank of the Allied positions provoked panic and dismay at Eighth Army HQ. Maybe the British forces had been over-committed? The Italian armoured forces now broke through, threatening the rear of the Allied defensive positions, and suddenly Gott's plan appeared to be unravelling. You felt the gloom growing once more around the headquarters staff. Gott threw his final reserves in to hold the Italian armour back, but this enabled the German panzers to seize the initiative, and by nightfall on 1 September the Eighth Army was reeling. Allied units began to fall back and only some localised counter-attacks managed to stem the tide, buying time for some order to be restored to the retreat. Gott's energy and determination drained away; he suddenly seemed exhausted. By the following morning he bowed to the inevitable and issued orders for the army to fall back to regroup.

Map 6: Allied Defeat at Alam El Halfa, August-September 1942

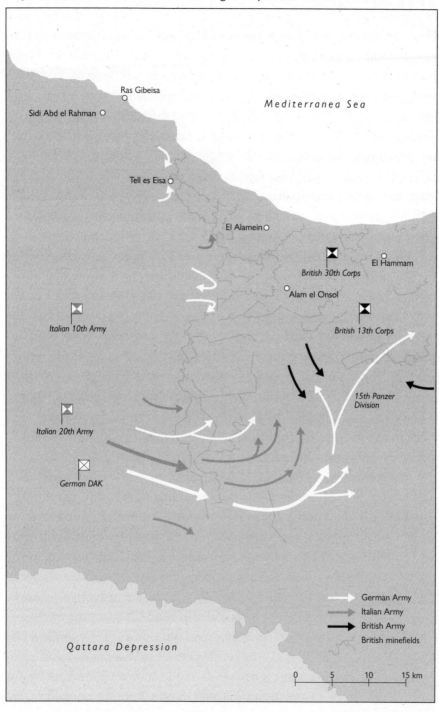

AFTERMATH

Panic broke out in Cairo as news of Rommel's initial success leaked out. British plans to evacuate now kicked in, even though the likelihood of the Axis forces reaching that far was still remote. Egyptian resistance groups and political opponents of British control in Cairo openly started to cause trouble, and General Alexander had to deploy troops and security forces to try and maintain order. Officials and administrative personnel began to prepare to withdraw. In London, Churchill raged, while in Washington President Roosevelt and his staff despaired at British military failings once more. Plans for the Allies' amphibious invasion of French North Africa, codenamed Torch, were now clearly in jeopardy.

Yet although Rommel had secured a great success, it was limited by his capacity to follow it up. The Eighth Army was once again in retreat, but it was neither shambolic nor without some degree of order; the Axis forces had themselves taken a battering and lacked the strength to pursue Gott's forces with anything other than tentative prods. Rommel despaired that the fuel and supplies he had been promised by Mussolini had failed to materialise, and his logistical lifeline stretching back to Libya was now quite simply at breaking point. There was only limited interest from Berlin, as Hitler was focused very much on the Eastern Front and the advance into southern Russia and the Caucasus – the campaign that would lead to Stalingrad.

As Gott's forces fell back to Alexandria and Cairo, Rommel's pursuit fizzled out. Although some light reconnaissance units reached the outskirts of Alexandria, there were simply too many Allied units to deal with. Gott regrouped some of his still-functioning mobile units and counter-attacked. Success was limited, but Rommel knew the game was up; his troops lacked supplies and his tanks had exhausted their fuel reserves. One thing the Allies had done effectively was to destroy anything that might have aided Rommel's advance.

With the advance stalled once again, just short of Alexandria and Cairo, Rommel fell ill and was recalled to Germany for rest and recuperation; he also planned to use the time to lobby Hitler directly for further resources to finish the job in Egypt. Churchill conceded that

Gott was not the man for the job and acted ruthlessly. General Harold Alexander took up direct command of the Eighth Army, pulling together the defence of Egypt while preparing a counter-offensive. On 15 October the Eighth Army attacked and the German and Italian forces fell back to the El Alamein position once more. Rommel returned to oversee the defence and explore options for returning to the attack, although little had been forthcoming from Hitler, who was beguiled by the developing battle at Stalingrad. All Rommel's hopes ended when news came through in early November of Allied landings in Morocco and Algeria, both parts of Vichy French North Africa, the loss of which threatened his supply lines, which still stretched back to Tripoli in Libya. Once the Allies pushed from Algeria into Tunisia, it was only a short hop to Tripoli and then Rommel would be cut off or dependent on the smaller ports of Tobruk and Benghazi in eastern Libya, which were always vulnerable to the Royal Navy. Realistically Rommel had no option but to retreat, but Hitler refused to allow it. Within a few months disaster had overtaken the Axis forces in North Africa.

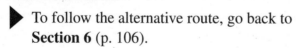

THE DECISION

▶ To follow the alternative route, go back to **Section 6** (p. 106).

OR

▶ To explore the decisions of the Mediterranean campaign, see the **Historical Note** below.

HISTORICAL NOTE

Richard O'Connor leading an advance into Tripoli in the spring of 1941 (from Section 1, go to Section 2) is one of the most fascinating 'what ifs' of the Second World War. If the British-led forces had delivered such a *coup de grâce* to Mussolini's African Empire, the impact might well have been pivotal – it would surely have reshaped the war. No need for El Alamein, quite possibly no one would have ever heard of Montgomery, and the Anglo-American invasion of Vichy French North Africa in November 1942 would have been rendered unnecessary. And of course Weygand's attitude to the Vichy regime might well have changed radically, with the British in Tripoli and on the Tunisian border. If the Germans had forced through what were historically referred to as the Paris Protocols in response to the Italians losing Libya, Weygand might well have switched sides, which would have unhinged Vichy France completely (from Section 2, go to Section 5). Such a commitment would obviously have had profound consequences for British policy in the Balkans and the Eastern Mediterranean – potentially ruinous implications.

Yet there were many obstacles confronting O'Connor in February 1941, most obviously some near-insurmountable logistical hurdles. The British advance had almost run its course around the time of the destruction of the Italian Tenth Army, so to press on to Tripoli was pushing the envelope and very high-risk indeed. In later life O'Connor always felt it a great missed opportunity, but others have questioned whether it was ever really feasible. Even if he had, Britain's major concern at the time – spring of 1941 – was still the Balkans. Churchill halted O'Connor and Wavell in February 1941 precisely because it was vital to British interests to keep the Axis powers out of the Eastern Mediterranean; this meant acting in the Balkans, first in supporting the coup in Yugoslavia and then in sending troops to Greece to support it against the Italians and then the Germans.

If the British had hesitated in committing resources to the Balkans, it could have caused Turkey to waver from its neutral stance. This is the possibility explored by going from Section 2 to Section 4. Even if Weygand had stayed resolutely neutral and the British had halted at Tripoli, it would still have had consequences in the Balkans, and particularly in Turkey. Yet it is unlikely that President İnönü's government would have fully embraced one side over the other, especially with the Germans on their borders in the west and the Soviets in the east. Nonetheless, it was always a major risk in committing further resources to capture Tripoli and even more so in pushing into Tunisia, when it risked Turkey being drawn into the Axis sphere of influence. This would surely have been calamitous for British control of the Middle East.

The historical route therefore (Section 1 to Section 3) made wider political sense. It of course ended in disaster in Greece, but it made some sense to show commitment and intent. Ultimately this path resulted in the need to fight at El Alamein. Here the choice lay between a popular, experienced hand in Gott and a new broom in Montgomery. Brooke may well have tried to get Monty installed, but Churchill forced through Gott (Winston never really liked Monty), so the historical route (Section 3 to Section 6) only leads to Monty after Gott gets killed. It is worth noting that Monty was by no means an obvious choice and that it was Brooke who saw in him the qualities to be successful.

But what if Gott had commanded at Alam el Halfa (Section 3 to Section 7)? Montgomery reduced the chance of losing the battle precisely because he was much less likely to take risks. Rommel preyed on British generals who tried to play him at his own game – freeflowing manoeuvre. Gott might have reacted and counter-attacked, and indeed the only realistic way that Rommel could win at Alam el Halfa was if the British blundered. Yet even if they had, Rommel making further progress to Alexandria or Cairo was as much of a stretch for him in September 1942 as it had been for O'Connor in pressing on to Tripoli in February 1941.

Eric Dorman-Smith served in the British Army's North African campaigns until the summer of 1942, when he became a casualty of the cull that followed the retreat to El Alamein. In subsequent posts he failed to impress. In later life he retired to his family home in Ireland, changed his name to Eric Dorman O'Gowan and supported the IRA in the 1950s.

Jacques Tarbé de Saint-Hardouin served as a distinguished career diplomat for France into the post-war years, culminating in his appointment as Ambassador to Turkey in 1952.

Alan Brooke served with great distinction throughout the Second World War and was Churchill's primary source of military information and advice from 1941 until 1945. They often clashed, but their differing personalities created an excellent working relationship.

İsmet İnönü was President of Turkey from 1938 until 1950 and then later Prime Minister. He continued to play a major role, sometimes controversially, in Turkish politics until the 1970s.

Andrew Cunningham proved to be the Royal Navy's most successful and highly respected commander of the Second World War, leading at a number of key battles, such as Taranto and Cape Matapan. He later served as First Sea Lord, the official head of the Royal Navy.

Francis 'Freddie' de Guingand acted as Montgomery's Chief of Staff, a very demanding role as Monty defined it, throughout the rest of the war. Freddie's sociable and amiable personality created a safety net around his boss's awkward and difficult demeanour.

Charles Richardson remained as the Eighth Army's chief planner before serving in other roles, until Monty brought him back onto his staff for the invasion of Normandy in 1944. Richardson ended his military career as a general and later wrote a series of well-regarded military biographies.

3

STALIN'S WAR:
WAR ON THE EASTERN FRONT, 1941–45

SECTION 1
BARBAROSSA

1 July 1941, Kuntsevo Dacha (Stalin's personal retreat), outside Moscow

A convoy of military vehicles and cars snakes its way through a dense birch forest outside Moscow en route to Josef Stalin's secluded Kuntsevo Dacha. Hidden away from prying eyes, surrounded by high walls and guarded by secret police, the dacha is Stalin's personal retreat; it is from here that he has governed the Soviet Union with brutal and ruthless determination since he seized power in the aftermath of Lenin's death in 1924. Though he also directs affairs from his offices in the Kremlin, Stalin much prefers to work in his study in the dacha. It is here that he has laboured on many of the most far-reaching initiatives in Russian history: vast agricultural collectivisation, rapid industrialisation and the suppression of all potential and possible opposition. Indeed, since the late 1930s when Stalin decapitated the Soviet political and bureaucratic state to tighten his grip on absolute power, his authority has been thought unassailable.

Yet the column of vehicles approaching his dacha today is a visitation that Stalin probably views with great alarm; for the first time in years his position appears under threat. It is something you ponder with grave concern, for you are *Vyacheslav Molotov*, the Foreign Minister of the USSR, considered by some to be Stalin's heir apparent. You know that you are regarded as a rather colourless grey bureaucrat, not one to offer dynamic leadership; your colleagues have taunted you in the past by referring to you by the nickname 'Stone

1. Stalin with Vyacheslav Molotov

Arse', an epithet provided by none other than Lenin himself. Yet you are a shrewd and ruthless individual who never openly demurred when instructed by Stalin to sign the death-warrants of many colleagues during the crippling purges of the late 1930s. Even though you your- self were smuggled away by the secret police at one point in 1936, having given offence to Stalin, you were reprieved and by 1939 had emerged as Foreign Minister – and the crucial Molotov–Ribbentrop pact of 1939 that saw Poland carved up was down to you.

Now, however, everything is in a state of flux. You and your colleagues – Lavrentiy Beria (Secret Police), Kliment Voroshilov (Defence) and Anastas Mikoyan (Foreign Trade) – are arriving to make one of the key decisions in the history of the Soviet Union; on your choices will rest the whole future of the revolution and the fate of the people of the USSR, if not of the whole world.

How had it come to this? How could Josef Stalin – the iconic, fiercely determined and uncompromising leader of the Soviet Union, feared by everyone in the whole country – suddenly be under threat? Just over a week earlier all had seemed relatively well, until, on 22 June, Hitler's Third Reich had launched a massive invasion of the Soviet Union, codenamed Operation Barbarossa. German tanks, vehicles and troops poured over the border, inflicting crippling losses on the Red Army. Overhead the Luftwaffe dominated the skies, smashing Soviet aircraft on the ground and in the air. The USSR's forces seemed to have been taken completely by surprise, and chaos and panic broke out. German mobile columns were soon advancing with lightning speed across the Russian steppe, with little in the way of coordinated or determined opposition from your nation's forces; Hitler's prediction that 'We have only to kick in the door and the whole rotten structure will come falling down' seems startlingly prescient. Soviet losses in the first few days were so catastrophic that many in Moscow – Stalin included – simply could not believe them. Stalin seemed to be issuing fanciful and hopelessly misguided counter-attack orders to units and forces that had long since collapsed or been overrun. Within a few days well over 2,000 Soviet aircraft had been lost and hundreds of thousands of Soviet troops had been killed, captured or surrounded with no chance of rescue.

In the months leading up to Barbarossa, Stalin had seemed dismissive of mounting intelligence that Germany was indeed preparing to attack the Soviet Union. Richard Sorge, a German communist spy operating in Tokyo, had sent detailed copies of German documents to Moscow that outlined in great detail the plans for the invasion of the Soviet Union, including troop numbers and the start date. The intelligence was rejected: 'We doubt the veracity of your information,' Sorge was informed. Even Winston Churchill and Franklin Roosevelt alerted Moscow to the impending attack, but Stalin rejected this information too, as coming from hostile sources intent on starting a war between Germany and the Soviet Union for their own ends.

On 14 June two of Stalin's most senior generals, Zhukov and Timoshenko, had requested that the Soviet Union's armies be fully mobilised and put on a war footing, such was their concern at the mounting intelligence over a looming German invasion. Stalin roared back, 'That means war! Do you two understand that or not?' As far as you could tell, Stalin simply refused to believe that Hitler was intending to attack. None of you, including Stalin, were foolish enough to believe that the peace treaty with Germany would hold in the long term, but you all thought that your diplomacy and foreign policy had bought the Soviet

Union more time than you were to get. Even when a German soldier positioned on the German-Soviet border defected on the eve of Barbarossa and warned that the Germans were about to attack, he was ignored. In fact, Stalin had him executed for spreading misinformation.

Stalin was also more culpable for the poor state of senior leadership and command in the Soviet armed forces in June 1941 than was caused by mere strategic blundering. As part of his determination to eradicate all possible challenges to his leadership, he had embarked on a major programme of trials and executions in the late 1930s, a campaign that became known as 'The Terror'. Thousands of senior figures had been removed, tortured and put through a series of show trials, before being brutally executed. As far as you could tell, most of the executions had been merely preventative measures to terrorise others into conforming and to liquidate any possible opposition before it had formed. Many senior and well-respected military figures were also swept up in the trials and summarily executed, leaving gaps at the top of the Soviet army and air force. You and some of your colleagues are now wondering whether the failure of the Soviet Union's defences over the last few weeks is in some way connected to Stalin's purges.

With disaster unfolding before them as the German forces surged into Soviet territory, Stalin's generals had conferred on how to counter the invasion. They argued that Hitler would strike for Moscow and that the Soviet army should react accordingly. Stalin disagreed and told his commanders to head off a likely German offensive in the south to the resource-rich Ukraine. This proved to be another blunder on Stalin's part, piling further misery onto the USSR's already crumbling military. Stalin has been driving himself into the ground, sleeping little, and now he is looking ill – 'a bag of bones in a grey tunic' is one description.

On 28 June came the shocking news that Minsk had fallen, meaning that the Germans had advanced 300 miles already, and the invasion was less than a week old. More importantly, Soviet defences have been breached once again, leaving Smolensk and then Moscow hopelessly exposed. You watched as Stalin descended on his generals to berate them for their failure; even the resolute and determined Zhukov was reduced to tears. In a final despairing flurry, Stalin despondently announced that 'Lenin founded our state, and we've fucked it up.'

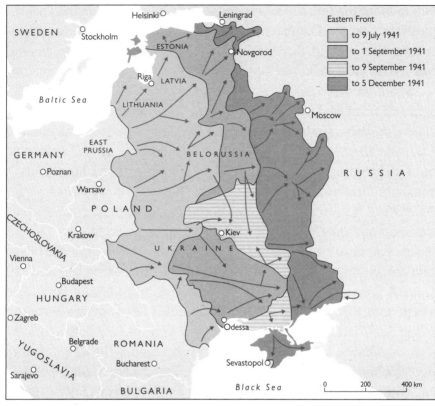

Map 1: Invasion of USSR in 1941

Then he fled to his dacha and seemed to retreat into isolation. In his country's greatest hour of need, Stalin brooded in his country residence and seemingly refused to engage with the affairs of state. Some wondered if he had suffered a nervous breakdown, though others whispered that he was still pulling the strings and that this was all a ruse to test the mettle of his subordinates.

What was to be done? The rest of the senior leadership group were stunned by Stalin's withdrawal. The whole structure of the state's command organisation was based on Stalin's vice-like grip; all decisions went through him, and his underlings rarely made any important moves without his agreement. In addition, a smell of fear hangs heavy in the atmosphere of senior government; each key player is wary of saying or acting in any way that another could report to Stalin as disloyal. The leadership is paralysed.

Now, for the first time, you have doubts about Stalin, and you are sure that your colleagues hold similar concerns, though no one is openly stating them yet. Stalin is clearly implicated in the disasters that are overwhelming the Soviet Union – it was Stalin who had dictated foreign and military policy, and he who had led them into this catastrophe. Is his time up?

If you make a move to replace Stalin, you have to be careful about Lavrentiy Beria, for as Stalin's security chief, he has always ruthlessly carried out Stalin's wishes and instilled fear and terror in those around him; one false move or statement in Beria's presence could spell doom. He has been feared and loathed in equal measure since becoming head of the NKVD, the state security organisation, in the late 1930s; you have heard that he murdered his previous boss soon after taking over. Stalin has overlooked Beria's manifest failings, including his well-known tendency to rapacious sexual predation; even his fellow Politburo colleagues, including Stalin himself, took special safety precautions with their female family members around Beria. He is too junior to take power himself, but can Beria be trusted at all?

More importantly, who would replace Stalin if you decide to eliminate him? Andrei Zhdanov is considered by some to be Stalin's natural successor. He has been a strong advocate of communist culture and ideology and is also a brutal oppressor, having been involved in the mass purges of the late 1930s. Zhdanov has, however, been sidelined by Beria in recent times and is now isolated in Leningrad, organising the defences of the Soviet Union's second city. Perhaps he is therefore out of the picture for the time being? Mikoyan is also held in high regard and may be a potential rival to you.

As your limousines draw up outside Stalin's dacha, you must decide upon your next move.

AIDE MEMOIRE

* Stalin has wielded all the power in the
  Soviet Union for years, however much of a
  liability he appears to be at this moment.
  Can you function without him?

* Stalin's egregious blunders have contributed significantly to this disastrous position. Maybe he has to go, and quickly?

* Stalin now seems to have thrown in the towel – or has he? He could simply be testing you. Maybe he is still manipulating the situation from his dacha?

* Stalin has exercised such a grip on power that removing him might actually create a power vacuum, as no one else knows how to take decisions and govern decisively.

* Would matters be improved if you remove Stalin? Could you, or anyone else, do better?

* The military position is catastrophic, and Stalin is to blame. Replacing him now with someone clear-headed, who is also resolute and ruthless, is surely the right move to save the USSR?

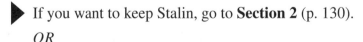

THE DECISION

What should you do? Consider the realistic options. Can, and should, you replace Stalin?

▶ If you want to keep Stalin, go to **Section 2** (p. 130).

OR

▶ If you want to replace Stalin, go to **Section 3** (p. 136).

SECTION 2
DEFENDING MOSCOW

0400, 17 October 1941, Kazansky railway station

It is a wet, very early morning in a darkened and chaotic central Moscow. All around, the city is in uproar, but at the Kazansky railway station a small group of high-profile staff are furtively bustling about, surrounded by a heavy security presence. In the midst of the activity on the platform, beside a readied armoured steam train, stands a short, squat, brooding figure, deep in thought. It is Josef Stalin, the leader of the USSR, a man in turmoil. He is contemplating the possible abandonment of Moscow to the invading Germans.

You are *Nikolai Vlasik*, Stalin's forty-five-year-old chief of personal security. It is a post you have held since 1931 and, after many years, you are now a trusted member of Stalin's close inner circle. The position has brought you a modicum of power, something you enjoy using when you can; it is a far cry from your humble peasant beginnings. You admire and fear Stalin in equal measure, but as you watch him pacing the railway platform you are gripped by the fear that maybe this time he is too hesitant.

It is a far cry from July, when Stalin rallied when confronted by his senior Politburo underlings at his dacha. Then you had sensed their fear and trepidation as they entered Stalin's residence, despite your boss's pensive, deflated and dishevelled state. Yet he was soon energised by what he heard. Molotov and his comrades had insisted that Stalin fully return to his duties, as he was needed now more than at any other time in the history of the Soviet Union. Molotov in particular

2. Vlasik in the late 1930s

had tried hardest to re-engage Stalin with the struggle. You later heard that Molotov had remained convinced Stalin had not been idle during his 'isolation' at all, and that this was a possible test of support. The reality was that as Stalin and his brutal henchmen had decapitated the upper echelons of government in the great purges of the late 1930s, there was no one willing or positioned to replace him. Cowed by years of repression, no one was willing to move against Stalin.

Within a few hours the Soviet Union's dictator was openly back in action, and although still clearly ill at ease with the deteriorating situation, he was able to speak, for the first time since the invasion, to the people of the Soviet Union on 3 July. Behind the scenes, you knew that Stalin had been desperately trying to open up a line of communication with Hitler in an effort to avert total calamity. His offers to trade all territory then under German control for a ceasefire and halt, unsurprisingly, fell on deaf ears in Berlin. Why would the Germans stall the invasion now, at the very moment when it seemed as though Stalin's regime might fall?

The German juggernaut seemed unstoppable. July and early August brought still further disasters for you all, as Axis troops, vehicles and aircraft inflicted stunning reverses on Soviet forces. Stalin's senior commanders looked on in utter despair and – despite the Germans changing tack for a time when they turned their attention to seizing resources and production centres in the Ukraine, which inflicted further crushing losses on your comrades – the Germans soon focused once more on capturing Moscow, the capital of Stalin's empire.

By 13 October the Germans had pushed to within a hundred miles of Moscow. Gaps were appearing in the Soviet defences, and German tanks and vehicles were sighted on the main route towards Moscow from the south. Panic erupted in the capital and martial law was imposed, in a vain attempt to maintain order. Looting and rioting broke out, and the roads were choked with traffic as citizens endeavoured to flee the capital before the Germans arrived. Stalin quickly conferred with General Georgy Zhukov, now leading the defence of Moscow. Zhukov was fiercely determined, outspoken, obstinate and argumentative, and Stalin, though jealous of Zhukov's military prowess, respected his bluntness and honesty – a real contrast with the fawning and obsequious Beria. Zhukov had temporarily fallen from grace in the first weeks of the invasion but had now rehabilitated himself. Could Moscow be held, Stalin demanded? Zhukov claimed it could, but that it would be a close call. Secretly you heard that his preferred strategy would be to pull back from Moscow and draw the Germans into grinding, attritional street fighting in the city, while holding better positions on the eastern outskirts.

You witnessed, however, that Stalin was being fed contrasting information by his acolytes. Clearly fearful for their own skins, Beria, Kaganovich and Malenkov, all senior party leaders, had advised Stalin to leave the city and reconvene the government in Kuibyshev,* situated on the River Volga to the east. Others were busying themselves with emergency plans, and the essential accoutrements of bureaucratic government were soon assembling at the city's railway stations, preparatory to evacuation. Armoured trains had been organised, convoys of trucks assembled and transport aircraft readied in anticipation. Rumours spread that Stalin was about to abandon the capital.

---

* The modern-day city of Samara.

Soviet evacuation card issued in the Second World War

At 0900 on 16 October Stalin had convened a meeting of his senior staff at the Kremlin to make a final choice about what to do. The advice from his jittery subordinates appears to have been to withdraw. They were fearful that to wait longer risked the leader being either isolated in Moscow or, worse, captured by the Germans. In truth, you knew they were also terrified that if they delayed longer, they too would be swept up in the chaos. 'How are we going to defend Moscow? We have absolutely nothing at all. We have been overwhelmed and we are being shot down like partridges,' Beria remarked darkly after the meeting.

Stalin, looking thinner and more haggard than ever, ordered the evacuation of the government; he announced that he would follow the next day, once adequate preparations had been made. By the evening Stalin's staff were heading towards the armoured train that was readying to carry the leader eastwards to safety. Yet although he had announced the evacuation, Stalin confessed to you that he was still unsure as to the wisdom of the move, and you saw him spend the day of 16 October ruminating and agonising about whether or not to run. By the time you arrived at the railway station in the small hours of 17 October, Stalin was still deep in thought. This could well be the most decisive turning point in the history of the Soviet Union.

AIDE-MEMOIRE

* It is distinctly possible that Stalin's evacuation would be the factor that causes the loss of Moscow.

* If Stalin stays in Moscow and fights on from there, there is every possibility that he could be trapped and caught up in the fighting, unable to lead the nation.

* If Stalin abandons Moscow and the Germans then capture the city, it could well spell the end of his regime.

* If Stalin remains, he might even be captured by the Nazis; imagine the damaging effects on the Soviet Union's chances of survival if Stalin was paraded through the streets of Berlin, prior to almost certain execution.

* If the Soviet Union collapses, Britain would be isolated once more, the Americans even less minded to intervene and, in effect, Germany would win the war.

* Stalin's flight from Moscow would be seen as a defeatist measure, one that indicated that even the mighty leader was now running for his life.

* Stalin has to survive to keep the Soviet Union going, and surely the best way of ensuring that is to relocate to a safer position to ensure that he is able to direct the war against the Nazis as best he can.

* Consider the effects on the population and, more importantly, on his fellow leaders if Stalin flees now.

* It is absolutely essential that Stalin survives to lead the nation; look at what had happened at the end of June, when he had appeared to step back. There is simply no alternative to Stalin moving to a place of safety.

THE DECISION

Stalin paces up and down the platform, pondering his next move; the armoured train is fired up, ready to carry him away. He turns to you for your view. What do you advise?

▶ If you suggest that Stalin relocates to safety, go to **Section 4** (p. 142).

OR

▶ If you advise Stalin to stay in Moscow, go to **Section 5** (p. 148).

SECTION 3
THE END OF THE REVOLUTION?

4 July 1941, the Kremlin, Moscow

A cavalcade of armoured cars and limousines sweeps into the Kremlin, closely protected by heavily armed military personnel and wary secret-service agents. Suspicion hangs heavily in the air and the tension is all too apparent. Hidden among the group of bodyguards and officials is Andrei Zhdanov, one of the principal figures in the Soviet Union's elite. He is a strong and resolute supporter of Stalin and a leading figure in the ideology of the Communist Party and, although a slightly lesser-known figure than Molotov, he is still a key member of the Politburo. Zhdanov is also keen to rebuild his reputation and standing, after being associated with the mismanaged Winter War against Finland of 1939–40 and the failed pact with Hitler, signed in 1939. Now he senses an opportunity. Zhdanov has taken great risks in flying to Moscow from Leningrad, where he has been preparing to defend the city against the rapidly advancing Nazis. Less than two weeks have passed since the Germans attacked, but it is already clear that the USSR is teetering on the brink of disaster.

You are *General Georgy Zhukov*, Chief of the General Staff, the man directly responsible for coordinating the defence

4. Zhdanov with Stalin

of the USSR against the Germans. A highly respected military figure of great standing in the Soviet Union's armed forces, with many years of experience, you have served in conflicts such as the First World War, the Civil War and more recently in the Far East. Following a message from Zhdanov, you have temporarily returned to central Moscow and accompanied him to the Kremlin to add your military backing to his visit. Like many others, you have heard that political control at the top of government has collapsed and that Stalin has ceased to function. Zhdanov has arrived to find out more and he needs your support to exert control and power.

On arrival, Zhdanov glowers darkly at all around him, and when Lavrentiy Beria, Stalin's enforcer, emerges, he is clearly taken aback at the presence of so many troops who will act on your say-so, not his. Beria's account of recent events fills you with dismay. When Beria, Molotov and their associates arrived at the Kuntsevo Dacha, just three days ago, Stalin had been sitting slumped at his desk, staring impassively into space, an apparently broken man. As the group entered, they had initially refused to make eye contact, so fearful were they of what they were about to do. Molotov had been the primary driving force in events, according to Beria. He declared that in view of Stalin's abandonment of his duties, they had to

5. Georgy Zhukov

take immediate and decisive action and appoint someone with credibility and standing to take charge of the situation. Stalin had glowered and sneered back that Molotov and his team were little more than 'yes' men, completely unable and unwilling to act decisively. Undaunted, Molotov carried on, declaring that Stalin's hapless mishandling of the strategic situation had imperilled the whole revolution and everything it stood for. Taken aback by the audacity of Molotov's pronouncement,

Stalin demanded to know on whose authority they were acting. Molotov brushed this aside and stated that he was assuming command, with Mikoyan and Voroshilov's support. Molotov then instructed Beria to place Stalin under immediate arrest. Stalin's protests were in vain, and he was dragged away by Beria's henchmen and bundled into a waiting limousine. As he was driven away, Stalin's bodyguards looked on in alarm and fear.

As Molotov, Voroshilov and Mikoyan headed off to seize control of the military situation, Beria sped to Moscow to the Lubyanka police headquarters, where he was now holding Stalin prisoner. On the morning of 2 July 1941 Beria had the charges against Stalin read out at a quickly convened and highly secret court, and then convicted the former dictator of betraying the revolution. Sentence was carried out at 1130, and a still-bewildered and disbelieving Stalin was executed by a gunshot to the back of the head, as so many of his victims had died. Beria quickly established that he had been acting under explicit orders from Molotov, but a cover story was issued to the world that Stalin had been killed in a German bombing raid while heroically defending the Motherland.

Matters had soon begun to engulf Molotov, however, and not only because of the remorseless advance of the German army, which appeared determined on seizing Moscow at the earliest opportunity. Molotov had been attempting to assemble the Politburo to organise a new command-and-control organisation for the defence of the Soviet Union when the news broke that Stalin was dead. Rumours spread that Molotov and Beria had carried out the great betrayal, but the suddenness and severity of their action caused shock – no one bought the cover story, of course. Beria feared that Molotov and his comrades would not be able to control the situation and he quickly distanced himself from Stalin's overthrow. He produced instructions from Molotov and Voroshilov that ordered the execution of Stalin.

Zhdanov is both appalled and galvanised by Molotov's act of gross treachery, little knowing that Stalin's resolve had apparently crumbled since the German invasion. You had clashed openly with Stalin a few days earlier at a meeting in which he called you 'useless', even though you had been carrying out plans imposed upon you. He had seemed unstable to you then.

At this crucial meeting at the Kremlin, you, Zhdanov and Beria now seize the initiative and, by controlling Moscow, are able to issue orders for the arrest of Molotov. Neither you nor Zhdanov trust Beria at all, but for the moment you need him on your side; you can deal with him later. An attempt by Beria's secret police to seize Molotov and Voroshilov was only partly successful, however, as Molotov was tipped off and retired south with some of his staff and other members of the Politburo. Defence Minister Mikoyan sided with you and Zhdanov, however, claiming that he knew nothing of orders to execute Stalin. During these days you watch on as panic and chaos breaks out in the USSR's government, with dysfunctional and fractured leadership during this critical moment in its history.

In the midst of this power struggle you still have to guide the Soviet armed forces as they try to resist the rampaging German military machine as best they can, but Hitler's troops advance remorselessly. As Zhdanov and Beria attempt to restore order, disaster after disaster threatens to engulf the Soviet Union. In the south, initially at Kiev, you heard news that Molotov and his faction have set up their own government, issuing orders that did not align with those of the Zhdanov coordinated regime in Moscow; both groups claim legitimacy, but neither could assume total power. Zhdanov has greater access to the levers of power in Moscow, but Molotov and his group have the higher-profile leaders. Molotov and his faction have even issued you, and many other army commanders, with differing orders than those coming from the Zhdanov government in Moscow; trust and faith in the political leadership are rapidly draining away.

AFTERMATH

Molotov's government was forced to flee Kiev in August, retreating to Kharkov, only to be threatened once again in the autumn by the remorselessly advancing German forces. By then Moscow was also in peril, and Zhdanov's regime was confronted by the agonising choice of whether to abandon the capital or fight on. You no longer believed that Moscow could be held, such was the chaos in the Soviet Union

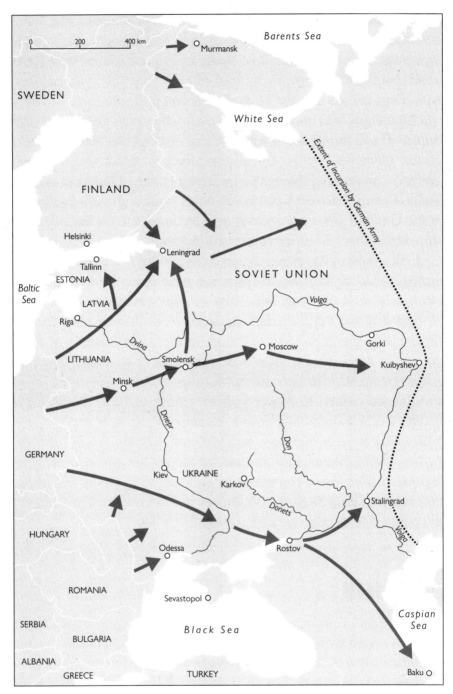

Map 2: Sketch map showing routes of German threats in the aftermath of the fall of Moscow

and the rate of the German advance. Your prognosis causes Beria to flee eastwards to Kuibyshev on the River Volga, soon to be followed by Zhdanov and the rest of the Soviet government. In the south, Molotov ironically fell back to Stalingrad, but in truth both factions had lost control of the defence of the Soviet Union.

By the end of October the USSR lies in ruins. Moscow has fallen, and in December Hitler begins issuing calls for the remaining elements of the Soviet government to come to terms. Some Soviet military leaders have even begun to negotiate their own deals. Differences between the two factions are being ruthlessly exploited by the Nazis, but neither faction is willing to surrender on anything like acceptable terms to Hitler. It will take yet more military campaigning by the Germans in 1942 to have any chance of delivering the *coup de grâce* to the USSR and bring an end to the revolution begun in 1917.

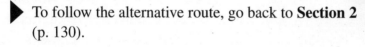

THE DECISION

▶ To follow the alternative route, go back to **Section 2** (p. 130).

OR

▶ To explore the history of the scenario, go to the **Historical Note** on p. 167.

SECTION 4
THE GREAT RETREAT

Early evening, 21 November 1941,
USSR State Defence Committee meeting,
the bunker, Kuibyshev

News of heavy military defeats is sweeping through Kuibyshev, once more undermining faith in the leadership of the USSR. Rostov has fallen that morning, which, along with the continuing occupation of Moscow, appears to have provoked a confrontation and possible showdown at that evening's State Defence Committee meeting. Even Molotov, a long-standing supporter of Stalin, now openly questions national strategy and policy.

You are **Anastas Mikoyan**, the forty-five-year-old State Defence Committee member responsible for supplying the armed forces with food and war materiel – an obviously crucial role at this time, and one to which you seem well suited. A long-time supporter of Stalin, you are also linked to introducing many Western consumer foods into the USSR, most notably ice-cream; as Stalin once joked, 'You, Anastas, seem more interested in ice-cream than communism.' You are renowned as a great political

6. Anastas Mikoyan

survivor, and you can now sense that the political tide is turning against Stalin.

The State Defence Committee meeting descends into recrimination and bitter argument, ending without much progress. In Kuibyshev a whispering campaign against Stalin's leadership has already begun, even within the Politburo. Has Stalin not been responsible for the collective failure to hold back the German invasion? Was it not his decision-making that has brought the Soviet Union to the brink?

In retrospect, Stalin's decision to abandon Moscow seems to have proved misguided. When, in October, Stalin and other key figures in the Soviet government fled the capital, it had undermined the resolve of those remaining to continue the fight. Moscow, the capital of Stalin's great empire, collapsed into panic and anarchy when he left, and not even the intimidating presence of his secret police could enforce discipline without the presence of the man of steel himself. Stalin's destination was Kuibyshev, a city on the Volga from where the war was now to be directed, in an effort to repel the advancing German invaders who were closing on Moscow at an alarming rate.

The military commanders left behind in Moscow now prepared to retreat, but not without extracting a heavy price from the Germans. General Georgy Zhukov had been tasked with the defence of Moscow; he had argued that it would have been possible, though difficult, to hold the capital. His preferred option, however, was to fall back, thus conserving available strength, preparatory to a planned counter-attack. His troops would draw the Germans into a ruinously costly battle of attrition on the outskirts of Moscow, before falling back to link up with fresh troops arriving from the east. Nonetheless, the loss of Moscow – even if it might only be temporary, as Zhukov planned and hoped – could still precipitate the collapse of the Soviet regime. Amid all the recriminations, would Stalin be able to cling to power?

The Politburo had slowly assembled at Kuibyshev, and by November Stalin was attempting to regroup and enforce his power and control once more. By then much had happened, however. German forces had pushed into Moscow on 15 November, seizing the Kremlin, the Lubyanka and Red Square. The city centre had fallen after sporadic fighting, and newsreels of panzers pulled up outside the Kremlin and

German troops parading through Red Square sent shockwaves through the Soviet leadership and the wider world. It mattered little that Zhukov's forces were still battling in the eastern suburbs of Moscow, or that the Red Army was assembling its reserves for a counter-attack. Military intelligence reports from the front recorded that the attack on Leningrad was stalling and that the German forces in Moscow were out on a limb. But would such crumbs of potentially positive news be enough for Stalin to shore up his support?

Stalin's colleagues plotted and schemed behind his back whilst Lavrentiy Beria, the head of the secret police, waited to see which way the wind would blow before throwing in his lot. Stalin still attempted to enforce his will, assuming that the fear he had generated during 'The Terror' of the late 1930s would be enough to keep him in power, along with Beria's scheming and brutality. Yet support for Stalin had been eroded by the death of Lazar Kaganovich, killed during the retreat from Moscow, and by Andrei Zhdanov's isolation in Leningrad; both men would have rallied to the cause. Now Stalin was dependent on Molotov and you to keep the faith, but others were beginning to break ranks, demanding change. With the Germans pushing remorselessly into Russia, some members of the Politburo were even worried that factions within the military might fall out of line and try to open negotiations with the enemy. Firm and clear leadership, untainted by continuous failure, was required now more than ever.

After the State Defence Committee meeting had broken up acrimoniously, you tried to assuage fears and worries. Your efforts are undone when, in the aftermath of the meeting, Stalin attempts to have Molotov and Voroshilov arrested. Beria, however, shrewdly noting that the wider leadership is beginning to turn against Stalin, informs the rest of the Politburo. You are all confronted by a stark choice: do you most fear Stalin, Beria's secret police or the Germans?

Everyone, including you, is also appalled that at this pivotal moment in Soviet history Stalin is resorting once more to extreme brutality and recrimination: this seals his fate, and Stalin is removed from office on 22 November 1941. Beria and his trusted policemen are despatched to deliver the news to Stalin, but Molotov and others still remain fearful that the whole situation might well be a test and that Beria's hitmen might return to arrest them instead. They make preparations with the military forces in Kuibyshev to contain Beria and his men, should it be required. The collective, if nervous will of the Politburo, however, convinces Beria that this time Stalin does not hold the winning hand, and he is placed under house arrest. His iconic status is still recognised as being important to national morale, however, and although Stalin's political power has been terminated, the rest of the Politburo decide to keep him in place as a figurehead for the time being. Stalin retreats to his summer house in Kuibyshev to enforced isolation, waiting for his long-term fate to be decided.

The following day the Politburo (with your support) presses Molotov to step up as leader; he is, in effect, the only readily available figure with enough gravitas to be taken seriously. He immediately draws upon the support of his fellow leaders in a more committee style of government. Molotov's position is soon bolstered by Zhukov's counter-offensive against the Germans in Moscow, which begins on 4 December.

Beguiled by the political capital to be gleaned from seizing Moscow when the Red Army had pulled back, Hitler had been drawn into the trap. Now, as Soviet forces rumbled forward, the Germans were forced to retreat and give up their great prize. German losses were not inconsiderable, and it was becoming patently clear that they were neither prepared for a winter campaign nor for a war lasting more than a few months.

Molotov's position is strongly reinforced by the success in winning back Moscow, and he and his staff quickly establish the narrative that the capital could have been defended all along, if only Stalin had remained resolute. Now Molotov, the firm, disciplined, professional leader, is in place to take control while the erstwhile dictator sits out the war in isolation, quietly drifting away into obscurity. It was later reported in 1944 that he had died of a heart attack. In 1942 and 1943 the Soviet army won spectacular victories at Molotovgrad on the Volga and then at Kursk, before a remorseless series of campaigns began to drive the Germans back to Berlin.

7. Molotov (right) confers with Roosevelt

AFTERMATH

Molotov instilled a more workman-like culture into the Soviet leadership, but one that nevertheless remained implacably opposed to the Nazis. Less driven by psychotic tendencies and darkness, but still wedded to brutality and indifference to human suffering, Molotov was able to lead the Soviet Union with some political skill throughout the rest of war. He of course remained suitably suspicious of the slippery British and arrogant Americans, but had a better working relationship with both Roosevelt and Churchill, leading to a more coordinated outcome to the post-war peace settlements, though not sufficiently so to avoid the descent into a Cold War. By 1945 the war was won, and the USSR under Molotov's bland but hard leadership had prevailed.

THE DECISION

▶ To follow the alternative route, go to **Section 5** (p. 148).

OR

▶ To explore the history of the scenario, go to the
Historical Note on p. 167.

SECTION 5
STALINGRAD GAMBIT

14 June 1943, Saltsjöbaden, Sweden

Some fifteen miles outside Stockholm, in the picturesque Swedish seaside resort of Saltsjöbaden, two small groups of foreign officials surreptitiously and furtively begin assembling in strictest secrecy at a secluded private house. One group is headed by Paul Schmidt, of the press and information section of the German Foreign Office, based in Sweden. He had been alerted a few days earlier that a senior Soviet figure passing through Sweden had briefed his staff to reopen peace talks with Berlin, however speculative and tentative. Schmidt had then been approached by Boris Yartsev, a senior figure in the Soviet embassy, and a meeting has quickly been organised. The Soviet group is headed by Mikhail Nikitin, a counsellor in the Soviet lega-tion in Stockholm, and although no senior ministers or high-ranking diplo-mats are yet involved, the Soviets seem to be genuine. There have been other secret discussions before, but the atmosphere in the Soviet camp now appears to have changed.

Yet all is not as it seems, for Boris Yartsev does not actually exist. You are ***Boris Rybkin***, a secret agent of

8. Boris Rybkin

the Soviet Union's NKVD, and you have been placed undercover in Stockholm as a diplomat under the name of Boris Yartsev. Previously you had been playing the same role in Helsinki, at the behest of Stalin himself, to try and negotiate an anti-Nazi deal with the Finns. You had then been moved to Stockholm and have been secretly guiding and overseeing back-channel links and communications with the Germans ever since, despite the ongoing war on the Eastern Front. In the first few months of the war the Germans had, unsurprisingly, largely blanked any approaches from Moscow, as the Nazis appeared to be on the brink of victory. That position has now changed, because the USSR is still fighting and the balance of power has equalised.

Stalin's decision to fight on in Moscow in the autumn of 1941 had been one of the main factors in halting the German assault. Zhukov's forces, backed by Stalin's iron will, Beria's brutality and the sacrifices of thousands of Soviet citizens, had combined with the deteriorating weather to stem the Nazi flow. The Soviet Union survived, and Hitler's troops were driven back, suffering appalling losses themselves in the vicious conditions of the Russian climate, something for which they were quite unprepared. By staying and facing down the Nazis while others, Beria in particular, were panicking and urging flight, Stalin had cemented his position as the iconic and resolute head of the Soviet Union; he was secure.

Nonetheless, 1942 had still been perilously difficult. The Germans had gone back on the offensive once again in the summer, pushing deeper and deeper into Soviet territory, this time into southern Russia. Hitler's target appeared to have been the oil fields of the Caucasus, which provided the lifeblood of the Soviet war machine. Stalin and his staff had apparently expected a German offensive against Moscow and so were initially wrong-footed by the enemy's strategy, so you heard. Partly as a consequence, the Soviet armed forces had suffered further heavy defeats and enormous loss of personnel and war materiel.

Yet the tide of the campaign began to turn. Part of the German strategy had been to push to the River Volga and capture the city of Stalingrad, in order to shield the flank of the main German advance to the oil fields in the south. But the Battle of Stalingrad took on a

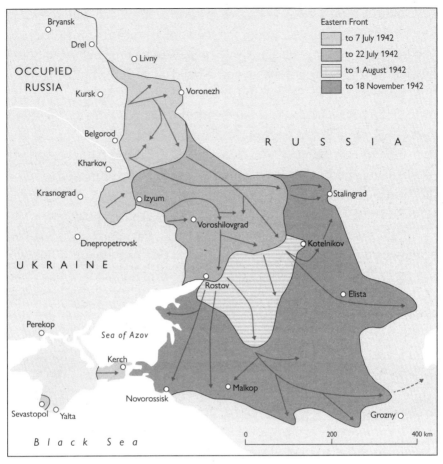

Map 3: German Southern Offensive, 1942

symbolic status and sucked in more and more German resources, diverting them away from other strategically critical and important missions. Over many weeks of grinding attritional slog, Hitler's forces managed to capture around 90 per cent of the city, but at a terrible cost in highly trained and valuable troops and equipment. The Soviet defenders suffered still greater losses, but Stalin was willing to pump them into the slaughter in order to tie down the Germans and draw their sting.

The Soviets were not simply taking the hit, however, but had prepared their own counter-offensive, which launched on 19 November. They soon surrounded the German Sixth Army fighting in Stalingrad.

Map 4: Soviet counter-offensive, 1942–43

Hitler seems to have refused to allow his generals to extricate the troops trapped in the city, arguing that they should be relieved by a new offensive, but this failed and by the turn of the year the Germans had been driven back across the entire front in the south of Russia and their forces compelled to abandon their hold on the Caucasus. The German army at Stalingrad surrendered on 2 February 1943, resulting in the loss to captivity of some 90,000 troops.*

It was a stunning victory for Stalin and the Soviet Union, even if delivered at a terrible cost; the Soviets lost more than one million troops, killed, captured, sick or wounded while fighting in the Battle of Stalingrad – considerably more than the Axis forces. But Stalin could more easily find fresh troops, and his industries were building

* Fewer than 6,000 survived the war to return home.

the replacements for equipment losses at a much faster rate than the factories in Germany. Yet although Stalingrad was a great success, the war still hung in the balance. Other Soviet offensives over the winter of 1942–3 had generally failed to make much progress and had sustained heavy losses, while in the spring of 1943 the Germans won a major victory around Kharkov, accounting for another 60,000 Soviet troops and 600 tanks. Hitler and his generals are surely now planning another offensive for the summer; rumour has it that it will be in the Kursk region.

By 1943, therefore, the war on the Eastern Front has in some ways reached a balance. Your Soviet forces seem to prevail in the winter, and the Germans in the summer. What is very clear to Stalin and his staff is the scale of the task confronting the USSR; to get the Germans completely out of the Soviet Union will require a vast investment and a huge sacrifice in lives and resources. By the spring of 1943 the longer-term prospects for the Soviet Union have improved considerably after Stalingrad, but victory is by no means assured and the cost of decisively defeating Germany will be heavy indeed.

Your instructions have shifted in tone over recent months; there is now no desperate need to strike a deal with Hitler, although communications should always be maintained, you are told by Moscow. Extracting an acceptable deal out of Berlin when the Nazis were in the ascendancy had been all but impossible, but by the spring of 1943, with the war reaching an apparent equilibrium in the east, maybe Hitler would be more amenable? The first tentative discussions between Moscow and Berlin in April had stalled, partly because the Germans had simultaneously announced to the world their discovery in the forests at Katyn in Russia of the mass graves of more than 20,000 Polish officers, murdered by the Russians in 1940, or so Josef Goebbels claimed. Moscow blamed Berlin for the brutal executions, but Soviet relations with the Polish government-in-exile in London fractured when an International Red Cross investigation, backed by the Poles, was agreed to by Berlin: Stalin proclaimed collusion and broke off diplomatic relations with the Poles.

Despite this new diplomatic hostility, tentative talks between Moscow and Berlin have continued in the background and only broke down because the Germans wanted an independent Ukraine, whereas the Soviets wanted a return to the 1941 border – something that even the Western powers had still not agreed to. The fate of these talks appeared to have been sealed when Stalin hardened his public position against Berlin in a major speech on 1 May, which underscored the unconditional-surrender policy of the Western Allies.

But on 4 June it was all once again thrown into the melting pot when Roosevelt and Churchill informed Stalin that the second front has indeed been postponed until May 1944. Stalin is furious, for it means that for another year the war will be won with Soviet blood while the cowardly Westerners skulk in England. To him, approaching the Germans to see how they might extricate themselves from the ruinous slaughter on the Eastern Front seems an appropriate response, even if one that he might not necessarily take. A major German offensive in central Russia is also looming large, so the matter has become quite pressing.

So Mikhail Nikitin and his delegation (with you keeping a careful eye on matters) have arrived in Saltsjöbaden with the intention of exploring the options with the Germans in a relatively open manner, unfettered by commitments to the Western Allies and driven by pressing strategic concerns. What might now be on the table? Were the Germans also willing to negotiate realistically?

The Soviets favoured establishing a buffer zone of friendly states on the Russo-German frontier along the lines of the 1941 borders, but Berlin still held out for an independent Ukraine. Hitler's team of negotiators believed they held the upper hand in the talks, as the German armed forces were ready to launch the big Kursk offensive in the next few weeks, but the chances of a complete victory against the Soviet Union still appeared remote. For the Soviets, the path to victory appeared long, arduous and costly, even if the Kursk battle could be won – they had excellent intelligence on the impending German attack. Was an independent Ukraine a price worth paying? Could a buffer state in Poland be established?

AIDE-MEMOIRE

* The Soviet Union appears to be on its own, now that the British and Americans are cowering across the English Channel waiting for Stalin's people to win the war for them.

* Why not strike a deal as a short-term solution and allow the Western powers to tackle Hitler for a while?

* Examine the map of Eastern Europe (Map 5) – demilitarised buffer states in Poland and an independent, demilitarised Ukraine might not be such a heavy price to pay for a reprieve in the fighting.

* If the Germans and the Western powers continued to fight, they would seriously weaken themselves while allowing the

Map 5: Proposed Border Regions

Soviet Union to rebuild and pull itself together.

* Hitler cannot be trusted.

* Any deal would surely be short-lived, and the long-term momentum of the war, which appears tentatively to be in the Soviet Union's favour, might well be lost if any deal is struck.

* It would be calamitous if, after Stalin had agreed a deal with Berlin, the USA and the UK also came to an agreement with Hitler. It would leave the USSR hopelessly exposed.

* It is more than apparent that Hitler's long-term grand vision is always for an empire based on expansion to the east, at the Soviet Union's expense. Accepting a deal now will surely merely push the final showdown into the future, when Germany might be better prepared.

THE DECISION

You have been instructed to convey the essence of the negotiations directly back to Stalin in Moscow. As you listen to the negotiations, you must decide what to advise. Should Stalin double-down on the rhetoric and continue the fight, or should he extract the Soviet Union from the war through a deal with Hitler?

▶ If you want to carry on fighting, go to **Section 6** (p. 156).

   OR

▶ If you want to do the deal, go to **Section 7** (p. 162).

SECTION 6
THE ROAD TO BERLIN

24 June 1945,
Red Square, Moscow

It is a rainy day in Moscow and the Soviet Union is celebrating its greatest-ever military victory: the crushing of Nazi Germany. In Moscow's Red Square, Stalin and the senior leadership of the Soviet armed forces watch on with smiles and pride as 40,000 troops and 1,850 military vehicles march by, in a triumphant parade that lasts over two hours. After four long years of total gruelling war, the Soviet Union has prevailed. Stalin revels in the success, a victory that has elevated his status to rival that of Lenin. Stalin has ordered the parade in order to mark his nation's greatest success (see Source 1).

You are **Colonel General Pavel Artemyev**, the forty-seven-year-old ex-political officer, born of a humble peasant background and now tasked with the responsibility of making the parade a great success. Fortunately as you look on, despite the rain, all goes well and Stalin appears pleased.

Yet the cost of the war has been enormous; the price in human life alone is close to thirty million people, although that will never

9. Pavel Artemyev

To mark the victory over Germany in the Great Patriotic War, I order a parade of troops of the Army, Navy and the Moscow Garrison, the Victory Parade, on June 24, 1945, at Moscow's Red Square.

Marching on parade shall be the combined regiments of all the fronts, a People's Commissariat of National Defence combined regiment, the Soviet Navy, military academies and schools, and troops of the Moscow Garrison and Military District.

My deputy, Marshal of the Soviet Union Georgy Zhukov, will be the parade inspector. Marshal Konstantin Rokossovsky will command the Victory Parade itself. I entrust to Col. Gen. Pavel Artemyev the preparations and the supervision of the parade organisation, due to his concurrent capacities as the Commanding General of the Moscow Military District and Commanding Officer in charge of the Moscow City Garrison.

June 22, 1945

MARSHAL OF THE SOVIET UNION JOSEF V. STALIN
Supreme Commander-in-Chief, Armed Forces of the USSR and concurrent People's Commissar of Defence of the USSR

Source 1: Order #370 of the Supreme Commander-in-Chief, Armed Forces of the USSR and concurrent People's Commissar of State for Defence

be admitted.* And the Soviet Union, although now backed by great military force, is weary and battle-scarred.

However, the success of 1945 is self-evident; the spectre of Nazism

* The true figure was shrouded in mystery and would remain a closely guarded secret for many decades to come.

and Hitler has been expunged and the Third Reich's collapse in 1944–5 has been precipitous, exposing its weaknesses and its lack of planning and management – things at which the Soviet Union has excelled. In retrospect, the notion of accepting a damaging deal with Hitler in June 1943, which in any case had only ever been a possibility rather than a probability, appears foolish and is quickly disavowed in the post-war writing of the Great Patriotic War. Any lingering chance of a deal being drawn up in June 1943 had in any case been shattered when the Swedish newspaper *Nya Dagligt Allehanda* (*New Daily Everywhere*) got hold of the story and published on 16 June that the Germans and Soviets were negotiating in 'a boarding house' outside Stockholm. The sensation was later reported in the *New York Times*. Both sides vehemently denied the veracity of the story the following day and retreated to their previous belligerent positions.

In truth, even as Stalin had pondered and then rejected the possibility of offering a serious deal to Hitler in June 1943, the tide of the war had in fact been turning against Germany far more than was imagined. The much-vaunted Kursk offensive in July 1943 – an operation that seemed likely to the German high command to deliver some degree of success – proved to be a dismal failure. The Soviet armed forces were well prepared and dug in, alerted by excellent intelligence and backed by deep reserves ready to launch their own counter-offensives to exploit the situation if the situation developed favourably. More than three million soldiers, 10,000 tanks and 4,500 aircraft fought for over six weeks in a bitter, attritional and unforgiving battle, and although the Soviets suffered heavier casualties, they prevailed, holding and then throwing back the German offensives, before going on the attack themselves. In the midst of the fighting around Kursk, Allied air and seaborne landings in Sicily also forced a redeployment of Axis units from Russia to the Mediterranean; Stalin's allies might not have offered the second front he wanted, but with their actions in Sicily, and increasingly with the bomber offensive against German cities and industries, they were contributing something meaningful.

After Kursk the Germans were never again able to mount a strategic offensive in the east. Germany's crippling losses could not be so easily replaced, whereas the Soviet Union had a greater capacity to

STOCKHOLM, Sweden, June 16 —Stockholm had a mild sensation today when the newspaper Nya Dagligt Allehanda announced in an extra edition that peace negotiations had been conducted recently "in a boarding house somewhere in Sweden" between Soviet representatives, among whom were Mme. Alexandra Kollontay, the Soviet Ambassador here, and high German officers. The negotiations reportedly came to naught over failure to agree on territorial questions, mainly concerning the Ukraine, which the Germans coveted but the Russians did not want to relinquish.

In informed diplomatic and political circles in the Swedish capital the report was treated as a post-dated April fool joke. Inclusion of Mme. Kollontay among the Soviet delegates was considered an automatic denial of the story, since the 71-year-old Soviet envoy still is convalescing from a brain hemorrhage and is authoritatively said to be in no physical condition to conduct negotiations of any kind.

Chargé d'Affaires Vladimir Semenoff is attending to Soviet interests in Sweden in her absence. Mme. Kollontay has not been in Stockholm for more than two months. Moreover, it is taken for granted here that Premier Joseph Stalin's May Day speech definitely shut the door on any peace talks with the Germans, who now more than ever would like a compromise settlement with Russia.

The newspaper said it had it from "a most reliable source" that Russia during these alleged talks had declared a willingness to cede the Baltic states and all Poland, while Germany wanted the Ukraine as well.

10. *New York Times* clipping, 16 June 1943

replace equipment and frontline troops. By the autumn of 1943 the Germans were falling back everywhere; all the territory Stalin had considered negotiating over was now being won by uncompromising force. With Hitler in retreat on the Eastern Front and in Italy, defeated in the Battle of the Atlantic and with German cities being pummelled from the air, the war had turned decisively in the Allies' favour.

Secured by his uncompromising and hard-line attitude and persona, Stalin was able to drive the USSR not only towards victory against Germany, but also to a position of long-term security. By appearing never to have flinched, Stalin was able to impose his will on foreign and defence policy ever more and enforce a vice-like grip on those nations in Eastern Europe that had aligned themselves with the Third Reich against the Soviet Union. Even Poland, supposedly an Allied nation, was to be brought to heel in 1944.

Churchill and Roosevelt looked on in dismay as the Red Army advanced across the Balkans and into Poland and Hungary, imposing pro-Moscow regimes, with an emboldened Stalin implacably driving them on. When Soviet troops battled their way into Berlin in April 1945, causing Hitler to commit suicide, Stalin's position as undisputed head of the USSR was reinforced still further. Even when, at the Potsdam Conference in July 1945, the new American President, Harry Truman, attempted to intimidate Stalin with news of the recently developed atomic bomb, the great dictator of the Soviet Union was unfazed, or so it seemed.

AFTERMATH

Diplomatic relations between East and West slowly deteriorated after the war, and Stalin's unassailable position in Moscow caused his domineering and near-psychotic personality to grip Soviet foreign policy in an implacable, yet destructive manner. Ultimately, Stalin's determined and uncompromising leadership in the war was later to prove a stumbling block to rapprochement with Washington, and led to the deepening of the Cold War. Hopes had been high in the USSR in 1945 that there might be a peaceful transition to the post-war world, where resources could be focused on rebuilding the nation. It was not to be, for the USSR was drawn into a costly and damaging Cold War with the USA and its allies, a confrontation that hindered its recovery and unduly absorbed its energies into the 1980s.

THE DECISION

▶ To follow the alternative route, go to **Section 7** (p. 162).

OR

▶ To explore the history of the scenario, go to the **Historical Note** on p. 167.

SECTION 7
THE TREATY OF STOCKHOLM

2 p.m., 12 July 1943, Chancellery House, Stockholm

In a rather stilted and awkward ceremony in the grand and opulent Chancellery House in Stockholm, Joachim von Ribbentrop, Hitler's Foreign Minister, and Vyacheslav Molotov, the Foreign Minister of the Soviet Union, have met to conclude a crucial new treaty. You are **Christian Günther**, Sweden's Foreign Minister, and are there to oversee the

11. Christian Günther

occasion and ensure its success. As a highly experienced foreign-affairs civil servant and minister, you know full well that it is crucial for Sweden's ongoing neutrality and security that the war between the Soviet Union and Germany is brought to an end. Berlin had pressured Stockholm back in 1941 to allow the transportation of German troops from Nazi-occupied Norway through Swedish territory into Finland to fight the Soviets, something that would made you a target for Stalin. With warring states all around, Sweden has had to tread very carefully and try not to enrage either side. Although you personally abhor the Nazis, you have been criticised for being too pragmatic and conciliatory in dealing with them, and too hostile to Moscow. In truth, you see Soviet-style communism as a great threat, but have no wish to see Sweden, at peace since 1814, embroiled in any war. You are keen to ensure that Ribbentrop and Molotov sign the agreement and bring stability back to the Baltic world.

As you and many other nervous Swedish dignitaries, and German and Soviet officials, look on, the two men complete the signing of a peace treaty that seems destined to change the course of the Second World War and global history. The document and underpinning agreement will, on the face of it, bring to an end the bloody war of annihilation that has devastated the Soviet Union for more than two years. Discussions between the two great powers had continued apace since the initial meetings in Sweden some weeks earlier. Then the initially tentative talks had gone sufficiently well for the negotiations to reach further up, to Hitler and Stalin. All had been undertaken in the strictest secrecy, notwithstanding a leak to the press in Stockholm that had quickly been suppressed.

Initially Hitler had been dismissive and uninterested in the talks, as he believed that his new Kursk offensive, Operation Citadel, would deliver such a crippling blow to the Soviet Union that Germany would be placed very much on the front foot for future operations, or placed in a position whereby they could compel Moscow to bend to Hitler's will. But a steady stream of intelligence reports emerged in Berlin that began to build a picture of deep defences, huge enemy reserves and strongly fortified enemy positions confronting the Germans in central Russia. Though Hitler had been mustering strength for the attack for some time, repeatedly delaying the start date in order to do so, it now seems as though Citadel could well be a risky venture. He has turned to the ongoing talks in Sweden with a renewed interest. Might he be able to squeeze enough out of Moscow to warrant suspending his military offensives in Russia? Germany's forces had suffered heavily in the Stalingrad campaign the previous winter and, with the Western Allies in the ascendancy in the Mediterranean, alongside increasingly heavy bombing attacks against German cities, might not a deal on reasonably favourable terms be worth considering?

It seemed that Stalin was offering an independent non-aligned Ukraine, and a series of demilitarised buffer zones and states along the 1941 Russo-German border as a way forward. Although Hitler's staff supported a deal (many had a firmer grasp of the enormous economic and industrial problems confronting the Third Reich in mid-1943), the Führer wavered. He feared that his war effort would lose momentum and the German people might drift back to their small, inward-looking civilian lives. The intelligence reports about Soviet strength were worrying (much of this had been deliberately fed to Berlin by the Soviets in an effort to force the Germans to the negotiating table), but to stop now might result in never having another chance to finish the war decisively in Germany's favour and thereby fulfil his grand vision of the Thousand-Year Reich.

Stalin and the State Defence Committee believed they had a strong long-term hand in the negotiations, but worried about the short term and whether the USSR could survive another Nazi onslaught; Soviet military performance had improved dramatically since the dark days of 1941, but 1942 and early 1943 had still witnessed crushing setbacks and defeats, too. Stalin was also wise enough to recognise that any deal was unlikely to endure while Hitler was still the driving force in Berlin. It was likely to be little more than a reprieve, but one that the Soviet Union would use far more productively than the Third Reich, Stalin estimated. He also considered that the Western Allies, despite their craven cowardice in refusing to establish the second front, would not throw in the towel, with Western Europe still under German occupation. The war would surely go on, further draining the resources of the Fascist states while the Soviet Union could rebuild.

Roosevelt and Churchill had now got wind of the talks and clamoured for Stalin to keep fighting and abandon the negotiations – had it only been May when he had effectively committed himself to the unconditional-surrender policy, they asked? More concessions and offers of support came Stalin's way from Washington and London, inducements that he found both amusing and pitiful.

For a few days in early July, in any case, it looked as though the talks would stall, as Hitler hesitated: Citadel might still go ahead as planned after all, even if it was delayed a few days more. He switched back to accepting the deal a few days later on 8 July, however, when the Allies began amphibious and airborne landings in Sicily; Germany's ally, Italy, now teetered on the brink of collapse and desperately needed propping up with troops and resources, forces that were still mustering on the Eastern Front for Citadel. If a deal were quickly signed, the Germans would be able to redeploy units to Italy to stop the Allies dead and keep Mussolini's regime afloat.

Hitler took the plunge and instructed Ribbentrop to press ahead with the negotiations. Through their respective embassies and staffs in Sweden, messages were exchanged between Berlin and Moscow, which opened up the door to what would be titled the Treaty of Stockholm. An immediate cessation of hostilities was declared on 10 July, to be followed by the signing of the agreement two days later. The Baltic states of Estonia, Latvia and Lithuania would be re-established as non-aligned neutral powers, with Finland also returning to its pre-Winter War of 1939–40 borders. In the south, Ukraine was to be declared independent, although as a neutral power, with all German and Soviet troops withdrawn.

In London the Polish government-in-exile raged when it was announced that Poland would remain under German and Soviet administration, as per the 1939 Russo-German agreement, even though the Soviet zone would remain demilitarised. Following the death of the Polish Prime Minister, General Sikorski, in a mysterious flying accident off Gibraltar just a few days earlier on 4 July, the Poles were now led by General Władysław Anders, a confirmed anti-Stalinist determined to block any rapprochement with Moscow. But any hopes the new Polish leader had of salvaging something from the war now seemed to have been extinguished. The many Poles fighting with the Western Allies sank into despair.

Churchill was sympathetic, but he and Roosevelt were now confronting the reality of redefining Grand Alliance strategy in a world where the Soviet armed forces were no longer pinning down the German army in Russia. A direct and costly confrontation with the massed German army in Western Europe now seemed unavoidable. Behind the scenes Roosevelt and his staff were furious that Britain's refusal to launch an invasion of northern France in 1943, which had so infuriated Stalin, had precipitated this crisis. The Americans became reconciled to planning and building for a much longer war in Europe than they had previously anticipated and their attention began to be drawn to the Pacific.

Yet both Roosevelt and Churchill still held out hope that the Russo-German Treaty of Stockholm would prove to be transitory; they still worked on rebuilding diplomatic links with Moscow, in an effort to convince Stalin that the deal with Hitler was near-worthless and that they should continue to cooperate. Nevertheless, in Washington and London the road to victory against Germany, Italy and Japan now looked decidedly more tortuous.

AFTERMATH

Mussolini's regime still collapsed in September when Allied forces invaded, but over the autumn and winter of 1943–4 German troops transferred in significant numbers from Eastern Europe to Italy to halt the advance of the Western Allies, while Hitler began investing heavily in fortifying France. Over Germany the Allied bombing offensive stalled in the face of increased Luftwaffe opposition, reinforced by aircraft freed from fighting the Soviets. Allied planners looked on pessimistically at the enormity of the task now confronting them in Western Europe – the prospective cross-Channel invasion was on the brink of further postponement, if not cancellation.

Stalin spent the breathing space created by the Treaty of Stockholm to rebuild and reinforce; there was no relenting in Moscow. He and Molotov were well aware that Hitler was not to be trusted and that further conflict was likely – the Soviet armed forces had to be ready this time. The collapse of the Stockholm agreement was not long in coming, and in March 1944 an outbreak of fighting in Estonia between a pro-Soviet faction headed by Johannes Vares and a newly formed nationalist government headed by Jüri Uluots, which the communists claimed was being manipulated by Berlin, caused confrontation between Berlin and Moscow. The new Estonian government deployed troops on the streets of Tallinn and began rounding up left-wing opposition leaders, some of whom, it quickly transpired, had been murdered. It later emerged that British agents had been operating in Estonia in an effort to cause friction, but it is unclear how much of a role they played in provoking the confrontation. Bitter fighting soon broke out, and Stalin ordered his troops across the border to re-establish law and order and prevent an openly hostile government being established in Estonia. Berlin retaliated by sending ships to Estonia to show their support to the 'legitimate' government. Nationalist governments in Latvia and Lithuania, still smarting from the Soviet invasion back in 1940, threw their support behind Estonia's government, and the situation escalated very quickly. German and Soviet troops were soon clashing again and by the end of April, with both Berlin and Moscow blaming each other for the breakdown, open warfare on the Eastern Front returned.

As Germany and the Soviet Union returned to war, Roosevelt and Churchill breathed sighs of relief; D-Day – the invasion of northern France – was back on and a road to victory was reappearing, even if it seemed more costly and difficult now than it had seemed before Molotov and Ribbentrop met in Stockholm less than a year before.

THE DECISION

▶ To follow the alternative route, go back to **Section 6** (p. 156).

OR

▶ To explore the history of the scenario, see the **Historical Note** below.

HISTORICAL NOTE

This chapter is a story shrouded in mystery and controversy – secret meetings that may or may not have taken place, contested events and reinterpreted national stories. The historical path is that Stalin gripped power once more in July 1941 (from Section 1, go to Section 2), having been roused from his dacha and reinvigorated by the call of his senior staff. He then decided at the last minute to remain in Moscow in October 1941 (from Section 2, go to Section 5). Finally, and despite his fury at the Western powers' unwillingness to start the second front in 1943, he was unable (and unwilling) to get close to a ceasefire with the Germans in July 1943 and so the war rumbled on, turning decisively in favour of the Soviet Union, culminating in the victory parade in June 1945 (from Section 5, go to Section 6).

Yet the events, let alone the decisions, that precipitated this path are all hotly contested. Although the story of the visitation of Molotov, Beria and others to Stalin's dacha in mid-1941 have been much repeated, as indeed has the idea that Stalin was fearful for his future on the arrival of his colleagues, some historians have also argued that even if Stalin had withdrawn (though some reject even this), he was still manipulating the situation and controlling events during his retreat. It does seem as though he suffered an understandable dip in morale and determination for a time. Yet the consequences of Stalin's removal in mid-1941 would surely merely have added to the crisis; his state simply had no recent experience of operating decisively without him. It is quite possible that his death would have caused a bitter fight for the succession in the middle of the greatest crisis in the history of the USSR (from Section 1, go to Section 3). It is unlikely to have ended well.

The story of Stalin's momentous decision on the railway platform in October 1941 as the Germans closed on Moscow is also contested; it does seem somewhat fanciful and may well have been a propaganda construction to illustrate his fearless determination and his pivotal role in the Great Patriotic War. Yet there was clearly a key decision to be made: stay and risk being caught up in the fighting, or withdraw to ensure control of the government of the USSR from safety. Although the historical path was to stay, it was by no means so obviously the right choice in 1941; abandoning Moscow to save the nation had, after all, been the path against Napoleon. Yet in a political system where Stalin played such a crucial, dominant role, for him to show weakness by withdrawing to Kuibyshev might well have precipitated a crisis in confidence in his leadership and made him vulnerable. With other calamities still to come, his removal was a possibility, if highly unlikely (from Section 2, go to Section 4).

Would Molotov have made a better job of leading the USSR? He would almost certainly have resorted to harsh and brutal methods to win the war and secure his power base, but perhaps not with the same vengeful and arguably psychotic tendencies of his predecessor. It is still more than likely that a Cold War confrontation with the West would have followed victory against Hitler in 1945, despite Molotov's superior diplomatic skills; such a collision of ideologies was perhaps an inevitability. It is also worth noting that historically Molotov lived on until 1986, despite a series of heart attacks, and so would have outlasted Eisenhower, Kennedy and Nixon, and might even locked horns with Ronald Reagan and Margaret Thatcher.

Was a deal between the USSR and Germany a possibility in mid-1943 (from Section 5, go to Section 7)? Possible, but again unlikely. Roosevelt and Churchill knew there was a risk in delaying the opening of the second front, but reckoned that Stalin would not break up the Grand Alliance; working together was the most obvious way to win the war. Any deal that Stalin made risked fracturing the anti-Hitler alliance, and that would probably allow the Nazis off the hook. However annoyed Stalin was at the Western powers for postponing D-Day in north-west Europe until May 1944, the main stumbling block remained the unrealistic aspirations and attitudes of the Germans. Hitler remained convinced that Operation Citadel (the Kursk offensive) would be a success, and he was most unlikely to hand back any territory even if major concessions on the part of the USSR could be agreed. Nonetheless, a back-channel was open during this period, so a deal was always possible. Any agreement was most unlikely to hold for long, however.

Vyacheslav Molotov – forever associated with the infamous Molotov-cocktail improvised bomb – remained a key figure in the Soviet Union until 1949, when he fell from grace and was sidelined by Stalin. He returned to prominence in 1953, but his opposition to the de-Stalinisation policies of Nikita Khrushchev caused him once again to be excluded from power by the early 1960s. Despite ill health, he lived on into the 1980s.

Nikolai Vlasik was another stalwart supporter of Stalin until, in 1952, he too was abandoned. He was convicted of various crimes and sent to prison, and although he was pardoned in 1956, he was nonetheless considered a pariah. It took until 2000 for his name to be completely cleared, by which time Vlasik had been dead for more than thirty years. He was nevertheless considered important enough to be the central figure in a major television series in 2015.

Georgy Zhukov, after being sacked as Chief of Staff in 1941, rapidly rebuilt his reputation and was the mastermind behind many key operations and battles that brought about the defeat of Germany. By the war's end he was the most high-profile figure in the Soviet military, and this caused him to be another who endured the unfathomable wrath of Stalin. Zhukov was marginalised to unimportant roles and, despite a return to power under Khrushchev, he was excluded again in 1957, retiring to write his memoirs.

Anastas Mikoyan, like Stalin, was not Russian, hailing from Armenia, although his close links with the oppressive regime in Moscow did him no favours in his homeland. Unlike most of his contemporaries, Mikoyan survived the purges and various political realignments to endure almost unscathed until a peaceful retirement in the 1960s after the fall of Khrushchev. As a colleague put it, 'The rascal [Mikoyan] was able to walk through Red Square on a rainy day without an umbrella without getting wet. He could dodge the rain drops.'

Boris Rybkin (or Boris Yartsev) continued to work for the USSR's state security police (in various versions) and was present at the important Yalta Conference in February 1945. He died in a car crash in 1947.

Pavel Artemyev rose from obscurity to the rank of Colonel General heading the Moscow military district until 1947. His association with Beria caused him to be banished from Moscow in 1953 when Beria was executed, and Artemyev drifted into obscurity, dying in 1979, aged eighty-one.

Christian Günther's reputation eventually came under some scrutiny in the post-war years, with some critics arguing that his 'realist' approach to foreign policy steered Sweden too close to Nazi Germany, even if out of a policy of self-preservation. After serving as an 'apolitical' Foreign Minister, Gunther returned to his diplomatic career in the late 1940s.

4

MIDWAY:
DECISION IN THE PACIFIC, JUNE 1942

SECTION 1
NIMITZ'S DECISION

22 May 1942, US Pacific Fleet HQ, Oahu, Hawaiian Islands

The US Navy's Pacific Fleet HQ is alive with activity and fraught with speculation. The buzz among the senior staff has been growing for some weeks as information, orders and plans circulate in anticipation of a new showdown with Imperial Japan's navy. You are *Captain Lynde D. McCormick*, chief planning officer to the US Navy's Pacific Fleet Commander-in-Chief, Admiral Chester Nimitz.

1. Captain Lynde McCormick

You come from an American naval family; your father had been a Rear Admiral and had served in the Spanish-American War of 1898–1900. Highly regarded by your colleagues, you are considered to be correct and reserved, 'a smart clean-cut gentleman'. Already based at Pearl Harbor, the US Navy's Pacific Fleet HQ in the Hawaiian Islands, at the time of the Japanese navy's surprise attack back in December 1941, you had survived the cull of senior officers in the wake of the disaster, and in April 1942 you were promoted to head

Nimitz's planning staff. He relies on you for advice and insight, and you have been closely involved with daily briefings, planning meetings and in-depth discussions with Nimitz and all his staff as you all wrestle with the task of identifying the enemy's next move and how to counter it. Today brings an important development – a new intelligence report – and your comments could well shape a crucial decision that Nimitz will have to make. Over the weeks you have been advising him, you have noticed that the one man who appears to have direct access to Nimitz at all times is Commander Edwin Layton, the Combat Intelligence Officer for the US Navy in the Pacific, and he has just arrived at Nimitz's offices, along with his curious and rather idiosyncratic head of cryptanalysis, Captain Joe Rochefort. The ensuing discussion might well determine the flow of the Pacific War and help turn the tide against the Japanese onslaught.

The war in the Pacific against Imperial Japan had not being going at all well, ever since they had launched their surprise attack against the US Pacific Fleet at Pearl Harbor on 7 December 1941 – a date that would live in infamy, as President Roosevelt put it. In ninety minutes 350 of the Imperial Japanese Navy's aircraft, flown by the most highly trained elite air crew in the world, had put much of the US battle fleet in the Pacific out of action. Ever since, the Allies had been thrown into headlong retreat across much of the Pacific and Asia. From the borders of India to the northern coast of Australia and right across to the Central Pacific, the Allies have been on the defensive (see Map 1).

Yet all is not lost. The bulk of the Japanese Army is still tied down fighting a war that it seems unable to win in China, and the USA is confident that it has the industrial capacity to overwhelm Japan in the long term. Crucially the last main hope of holding back further Japanese expansion – the US Navy's aircraft-carrier fleet – escaped destruction at Pearl Harbor. Aircraft carriers have become the new capital ships of the Pacific War. Each big fleet aircraft carrier is equipped with sixty to ninety aeroplanes, usually a mix of fighters, dive bombers and torpedo bombers. These carrier-based air groups can project hitting power over a range of some 150–200 miles and, as had been demonstrated at Pearl Harbor, can sink even the mightiest of armour-clad big-gun battleships. The fast-moving aircraft carriers

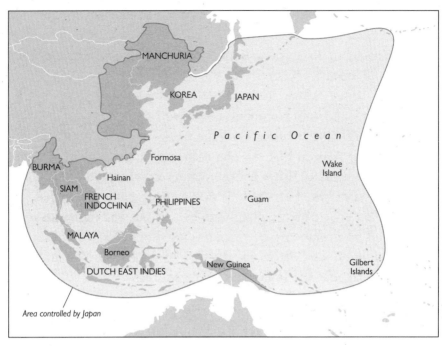

Map 1: Japanese expansion by mid-1942

with their long-range hitting power are surely dominant now. Although the Japanese have more carriers and better-trained crews in mid-1942, the Americans are still competitive; the US Pacific Fleet's carriers remain a lingering threat to the Japanese Navy.

To emphasise the point, in early May 1942 Japanese and American carriers battled against each other in the Coral Sea, east of New Guinea, without ever sighting each other; all the attacking had been conducted through air attacks. It was a scrappy encounter, but for the US Navy it at least demonstrated that they are not entirely impotent in the face of Japan's larger navy. In the wake of the Battle of the Coral Sea, Admiral Nimitz and his staff began to assess where the enemy might strike next. Nimitz (see Picture 3) is a Texan of German stock, and was appointed by President Roosevelt in the aftermath of the Pearl Harbor catastrophe with the rather broad, though straightforward directive: 'Tell Nimitz to get the hell out to Pearl and stay there until the war is won.'

Convivial in character and known for his amusing anecdotes,

2. USS *Enterprise*

Nimitz is well regarded for his intelligence and organisation, while being open to input and debate from all in attendance at his meetings, however junior; he is no dictator. He arrived in Honolulu to rescue the US Pacific Fleet from the gloom of the disaster of the Pearl Harbor attack. As one staff officer has put it, it was 'like being in a stuffy room and having someone open the window and let in a breath of fresh air'.

But the strategic position in May 1942 is still grim,

3. Admiral Ray Spruance (left) with admiral Chester Nimitz (right)

and Nimitz urgently needs a success to stem the tide of Japanese expansion. Even as the Coral Sea battle is playing out, it is abundantly clear that Japan is planning a new major offensive – but where (see Map 2)?

Intelligence chiefs in Washington are dithering and differing in

Map 2: Japan's strategic choices, May 1942

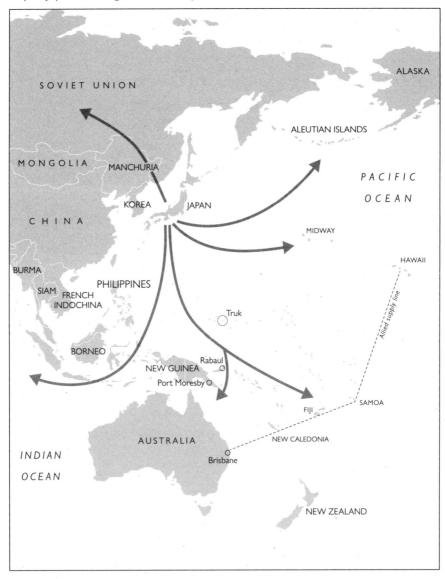

their assessments – some believe the Japanese will strike in the far north at the Aleutian Islands (part of Alaska); some that the South Pacific will be the focus; others that Hawaii could well be attacked again. The US Army has even hinted that raids on San Francisco or the Panama Canal are realistic possibilities. Confusion reigns supreme, and traditional intelligence sources are proving of little help.

Nimitz, however, has become increasingly appreciative of his own intelligence staff in Honolulu. At the start of the year they had been in the doldrums after their failure to spot the Pearl Harbor raid, but they recently correctly identified Japanese intentions prior to the Coral Sea battle. Nimitz appears to be particularly impressed by the thirty-eight-year-old Layton, who is considered a 'sharp, quick thinking, fast acting . . . human dynamo'. Nimitz even appears to have come round to appreciating the signals intelligence unit, or 'Hypo' as it is known. Signals intelligence relies on intercepting the enemy's coded or encrypted radio transmissions and deciphering them; this is the basis of the work carried out at Bletchley Park in Britain, which broke Germany's Enigma system, for example. The US Navy's team at Honolulu is headed by the highly unconventional forty-one-year-old individualist Captain Joe Rochefort Jnr.

Rochefort had lied about his age to gain entry to the US Navy in 1918, but has worked his way through the service demonstrating a penchant for intelligence work, particularly cryptanalysis and code-breaking, which resulted in his appointment to head Hypo in 1941 (see Picture 4).

It is cutting-edge work, largely conducted in a windowless underground facility in Pearl Harbor, where Rochefort's hand-picked team of 'pasty faced crypto-analyst troglodytes' labour round the clock, 'sealed off from the rest of the world like a submarine'.

4. Josef Rochefort Jnr

Rochefort himself often wears slippers and a bathrobe with his uniform, and you have heard that he will go days without bathing, such is his single-minded determination. Hypo's recent success has been based on collaboration with other intelligence units, and between them they seem to breaking into the Japanese Navy's JN-25 code, a process that includes the use of IBM tabulating machines and piles of punch-cards, as well human guile and ingenuity.

On 8 May Nimitz was visited by Layton, who persuaded him to come to Hypo and see the mounting evidence that pointed to Midway Islands – some 1,300 miles north-west of Pearl Harbor, and a vital air base defending the approaches to Hawaii – being a mysterious target called 'AF' in ongoing Japanese messages. Nimitz was still sceptical and had sent you instead to assess the intelligence. It had all been set out for you on makeshift tables by Rochefort and his team, and to you it seemed quite convincing. But Nimitz has also been receiving contradictory advice from Washington, suggesting alternative targets and for the attack to be in mid-June, not early June, as Hypo is suggesting.

Then, just a few days ago, Rochefort and his team proposed a ruse to prove once and for all that Midway was AF, and therefore the intended target of Japan's next offensive. They suggested that a message be sent by secure underwater cable to the US base at Midway to announce in an open, uncoded radio message back to Pearl Harbor that Midway's desalination plant had failed, and that fresh water was going to be in short supply. Hopefully Japanese listening teams would intercept the message and pass the information on by radio to the units readying for attack. It would be crucial for any invading force to know that lack of fresh water was going to be an issue. This alert could then, with luck, be intercepted, decrypted and read by Hypo.

Now, on 22 May, Layton and Rochefort have arrived at Nimitz's HQ with the results of that stratagem. As Nimitz, you and the rest of the assembled staff listen intently, Layton and Rochefort show you all a decrypted message that was intercepted just a few hours ago (see Source 1).

KIMIHI [Naval Intelligence Tokyo] - The AF
air unit sent following radio message to
Comdt 14th District: 'AK' on 20th. 'Refer
this unit's report dated 19th, at the present
time we have only enough water for two weeks.
Please supply us immediately.'

Source 1: Translation of the intercept about AF's (or Midway's) lack of water

Objective AF could be short of fresh water and therefore units preparing to attack would need their own extra supply, the message states. Surely this proves, Rochefort urges, that AF is indeed Midway? Layton seems to agree, and Nimitz, who has anyway been coming round to the idea that Midway is the target, concurs.

But it is not that straightforward. Army intelligence is still arguing that the US Navy should prepare for a flexible response; if Hypo is wrong and Hawaii is the real target, for example, it could prove catastrophic. Washington intelligence staff are also unconvinced and have put forward the idea that the target might be Johnston Atoll or the Aleutian Islands in the far north. It is also true that the ease with which Hypo is breaking and reading Japanese encrypted messages is highly suspicious; the ruse about the lack of fresh water is simply too convenient, and the Japanese could not possibly have fallen so easily into such an obvious trap.

Nimitz faces a real dilemma. Whatever or wherever AF is, time is running out; there now seems little doubt that the Japanese will attack in the first few days of June. He has few resources to deploy against a much larger, highly experienced and tactically adroit enemy fleet, and a flawed American naval deployment based on overconfident intelligence assessments could prove disastrous. But if Rochefort and his team are correct and their assessments of enemy intentions are spot-on, a huge advantage would be conferred on Nimitz's admirals as they lead the US fleet into battle.

AIDE-MEMOIRE

* Think carefully about the trick played by Rochefort on the Japanese. Could the Japanese really have been so slapdash with their communications security?

* Hypo intelligence was broadly correct about the Battle of the Coral Sea.

* Deploying the US fleet to defend Midway could leave other targets, such as the much more valuable Pearl Harbor, open to attack.

* The only way to counterbalance Japan's much larger fleet is to catch it by surprise and get in the first blow.

* A flexible wait-and-see stance offers a greater chance of avoiding being caught out and taken by surprise. Might it be prudent to let the Japanese fleet show its hand before committing the US fleet?

* Striking a major blow against Japan's carrier fleet could dramatically rebalance the war in the Allies' favour.

* The US Navy has few enough assets as it is. Why risk them?

THE DECISION

What advice should you, as Nimitz's advisor, offer?

▶ To advise that Nimitz backs Rochefort's assessment and takes the risk of positioning the US fleet around Midway, to attempt to surprise the Japanese fleet, go to **Section 2** (p. 184).

OR

▶ To suggest adopting a flexible response, and to counter any clear moves by the Japanese fleet only when they have shown their hand, go to **Section 3** (p. 192).

SECTION 2
SPRUANCE'S LUCK

0615 (GMT -12 hours), 4 June 1942, USS *Enterprise*

It is daylight on the morning of what many on the bridge of the USS *Enterprise* – flagship of the US Navy's Task Force 16 under the command of Rear Admiral Ray Spruance – believe could be one of the most decisive and important days of the whole Pacific War. You are **Captain Miles Browning**, Spruance's Chief of Staff and Air Tactical Officer, known for being irascible, aggressive and hard-drinking. It is fair to say that you are respected and feared rather than liked, but those around you think you a shrewd and astute tactical officer with a sharp intellect, alongside a firm grasp of carrier air tactics. These skills are about to be called upon because Admiral Spruance, though highly regarded as a commanding officer, has no real experience of air operations. Spruance stepped up to command Task Force 16 only a few days earlier when your previous boss, the aggressive and outspoken Vice Admiral William 'Bull' Halsey, was forced to remain in Honolulu due to stress-related illnesses. Spruance is known as 'electric brain', and Halsey, while recognising that Spruance had no experience of commanding carriers, had recommended him based on his 'excellent judgement and quiet courage'. Spruance is quite the polar opposite of his former commander; he is a quiet and reserved, largely abstemious non-smoker who enjoys a morning mug of hot chocolate. You, like many others, are withholding judgement on him. But now, as the senior staff team assembled in the flag conference room on the USS *Enterprise*, Spruance quietly announces that crucial decisions have to be made very quickly.

Over the previous forty-five minutes or so a series of sighting reports from American PBY Catalina flying boats scouting north-west of Midway Island have been received. An apparently as-yet-unsuspecting Japanese carrier fleet lies a little under 200 miles to the west-south-west of Task Force 16 – your position. Much of its striking power appears to have been deployed against the island of Midway, leaving it potentially vulnerable to a concerted American air strike from the US carriers, if you can attack the enemy quickly. It appears that the intelligence

5. Captain Miles Browning

assessments of Layton and Rochefort back in Honolulu have proven to be spot-on after all. According to Layton and his team, Admiral Yamamoto, the Japanese Commander-in-Chief, was going to lead with his four big aircraft carriers, possibly in two groups, while his heavy battleship forces would follow on behind. Layton even predicted: 'They'll come in from the north-west on a bearing 325 degrees and they will be sighted at about 175 miles from Midway, and the time will be about 6 a.m.' How convincing was this intelligence, Nimitz enquired? According to Rochefort, they were so familiar with the Japanese fleet's workings they could even recognise the techniques of the Morse code operators on the various ships.

To counter this, the Americans have packed 140 aircraft onto Midway Island itself to offer stout resistance to any Japanese attack, although some of these aeroplanes are obsolete or unproven. Yet Midway is a fixed position and, even if better resourced, unlikely to completely catch out the Japanese; this could only be achieved by the mobile striking power of the US carriers, and it was here that the challenge of mustering sufficient striking power has been greater. At first it appeared that Nimitz had only two carriers ready to go: *Enterprise* and *Hornet*. *Saratoga* was still returning from a refit on the west coast of the USA, *Wasp* was in the Atlantic, and *Yorktown*, badly mauled at the Coral Sea battle, was only just limping back to Pearl Harbor. Initially engineers claimed it would take three months to repair the *Yorktown*, but Nimitz would have none of it: 'We must have this ship back in three days,' he had declared. Everyone had been impressed by the supreme effort made to patch up the *Yorktown* and, although less than 100 per cent ready for battle and carrying a rapidly thrown-together air group, she had been deployed for action.

As Spruance and Rear Admiral Frank Fletcher (commander of Task Force 17) headed out from Pearl Harbor, Nimitz issued them with a letter of instruction, based on the notion of calculated risk (see Source 2).

Sightings of Japanese transport ships slowly heading towards Midway from the west on 3 June had confirmed that the attack was on, and the American carriers had readied for action in the small hours of 4 June some 300 miles to the north-east of Midway. At first light Fletcher had despatched some of *Yorktown*'s scouting aircraft to the north; he appears to be concerned that the Japanese might have further ships as yet unaccounted for; some thought was given to the idea that the Japanese would operate two groups of two aircraft carriers, some distance apart.

At around 0530 sighting reports of the Japanese fleet around 200 miles away begin to filter in. Then, at just after 0600, a clear report identifying the position of two Japanese carriers is received. A few minutes later Spruance informs you that he has just received a message from Fletcher: 'Proceed south-westerly and attack enemy carriers as soon as definitely located. I will follow as soon as the planes are recovered.' Fletcher is technically the senior commander of the American naval forces, but he will hold back to recover his scouting aircraft and provide cover and backup to Spruance's carriers. There are still two enemy carriers as yet unaccounted for. Spruance is to close and attack.

But there is a problem. The American carriers will have to close the distance between the US fleet and the Japanese carriers before you can launch an air strike; currently the distance is too great for some of your aircraft to reach the target and have a chance of returning.

UNITED STATES PACIFIC FLEET
USS PENNSYLVANIA,
FLAGSHIP OF THE COMMANDER-IN-CHIEF

May 28, 1942

Serial 0114W

S E C R E T

From: Commander-in-Chief,
United States Pacific Fleet

To: Commander Striking Force
(Operation Plan 29-42)

Subject: Letter of Instruction

1. In carrying out the task assigned in Operation Plan 29-42 you will be governed by the principle of calculated risk, which you shall interpret to mean the avoidance of exposure of your force to attack by superior enemy forces without good prospect of inflicting, as a result of such exposure, greater damage to the enemy. This applies to the landing phase as well as during preliminary air attacks.

C. W. Nimitz

Source 2 Nimitz's Calcultaed Risk directive to Fletcher and Spruance, May 1942.

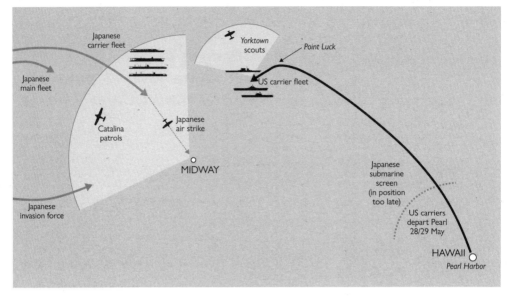

Map 3: The opening moves, morning of 4 June 1942.

Because of this, some of the staff team on *Enterprise* advise that the best time to launch a strike would be about 0900, close on three hours away. This launch point is dictated by the maximum operating range of your strike aircraft, particularly the Douglas TBD Devastator torpedo bombers, which can at best reach a distance of 175 miles before having to turn back.

Yet to you and your tactical air team, a great opportunity is apparent, if one replete with risk. You have reports that the Japanese carriers have launched an air strike against Midway to suppress its defences. These aircraft will eventually begin returning to their carriers and have to be recovered, to refuel and rearm, possibly for a second strike against Midway. In the middle of this recovery process the enemy carriers would be most vulnerable to an air strike. In fact this exact tactical situation is something you identified in a tactical thesis back in 1936. You quickly calculate that the Japanese carriers will be at their most vulnerable around 0930–1000; in order to attack during that window, your aircraft will have to begin launching a little after 0700, long before 0900. However, as you concede to the assembled staff group, this risks some of the aircraft running out of fuel before

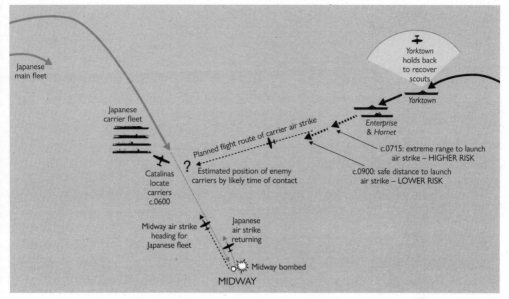

Map 4: Rapid or delayed launch, morning of 4 June 1942.

they can get back to the US fleet. You could reduce the return distance if the American carriers closed on the Japanese fleet, but that would leave Spruance no manoeuvring room. If he had to react to another threat and turn away from the Japanese carrier fleet he would risk heavy aircraft losses, as they would undoubtedly run out of fuel and then have to ditch in the Pacific Ocean, with the likely loss of many pilots and air crew.

What should you advise?

AIDE-MEMOIRE

* Consider Admiral Nimitz's calculated-risk principle (see Source 2 on p. 187). What approach best fits the Commander-in-Chief's instructions?

* Consider the intelligence gathered so far and the locations of the sightings (see Map 4). Is it compelling enough to go all

out on the attack immediately and take the gamble?

* An early launch could catch the Japanese out completely. It offers the best chance of a successful attack.

* Launching too early risks losing much of your air group - without them, you are completely open to counter-attack.

* As far as you are aware, the enemy has not yet located your fleet. You could get in the first blow before the Japanese can react properly.

* So far your reconnaissance aircraft have located only two enemy carriers. Where are the other two? Shouldn't you hold something back until they are located?

* The recent Battle of the Coral Sea seems to show that offence is more important than defence in carrier battles.

THE DECISION

Should you begin launching around 0715 and take the risk, or wait until closer to 0900 and ensure the survivability of your air group?

▶ To hold back and wait to reach the safer launch point of 0900, go to **Section 4** (p. 200).

OR

▶ To attack with everything as soon as possible at 0715 and hope to recover your aircraft later as best you can, and accept the losses, go to **Section 5** (p. 204).

SECTION 3
WASHINGTON'S FAILURE

0845 (GMT -12 hours), 3 June 1942,
USS *Yorktown*, c.300 miles
north-north-west of Pearl Harbor

It is early morning on the bridge of the USS *Yorktown*, flagship of Task Force 17 commanded by Rear Admiral Frank 'Black Jack' Fletcher, the man who had stymied the Japanese at the Battle of the Coral Sea just a few weeks ago. You are **Captain Elliott Buckmaster**, the fifty-two-year-old commander of the aircraft carrier USS *Yorktown*, a ship that you have captained since February 1941. Despite receiving substantial damage at the Coral Sea battle, *Yorktown* – under your careful direction – struggled back to Pearl Harbor to be patched up for the Midway battle. But you know, deep down, that the ship is barely fit for action.

6. Captain Elliott Buckmaster

As personnel bustle about, there is a clear sense of anticipation as the senior staff await possible news of Japanese activities. The American carriers – *Enterprise*, *Hornet* and *Yorktown* – have been positioned on station to the north of Pearl Harbor, ready to react to any offensive move made by the Japanese navy. You have all been briefed that intelligence sources in Honolulu and Washington have indicated that a Japanese attack is imminent, but that the location is contested. The Aleutian Islands in the far north, Johnston Atoll to the south, Midway to the west or even the Hawaiian Islands themselves are all possible targets. Admiral Nimitz strongly favoured Midway as the enemy's main objective, but Washington instructed him to accommodate other possibilities, most obviously a potential threat to Pearl Harbor.

The previous day, 2 June, you heard that Japanese ships had been sighted heading towards the Aleutian Islands to the south-west of Alaska, but Nimitz, back in Pearl Harbor, told Fletcher to await further news. Then, a little after 0845 on 3 June, news breaks that a Midway-based Catalina reconnaissance flying boat has detected a Japanese transport fleet some 700 miles west-south-west of the island – it seems that Midway is indeed to be the objective of the enemy's offensive after all. Fletcher curses and fumes; Nimitz and the intelligence team at Honolulu had been right. But because of the caution imposed by Washington, the American carriers are located too far away to react quickly to the events unfolding at Midway. Shortly afterwards you receive a new signal from Nimitz (see Source 3 on p. 194).

Fletcher instructs you to begin immediate preparations for a move westwards at best speed towards Midway. But it will take the best part of two days to reach a position from which to intervene; you all realise the chance of surprising the Japanese fleet has been squandered.

The following morning, 4 June, Catalina flying boats begin signalling that they have located the Japanese fleet – comprising four large aircraft carriers, along with escorts – some 175 miles north-west of Midway. Rochefort's intelligence assessments were proving to be unerringly accurate. Later that morning you receive updates from Midway itself that it has been heavily attacked by enemy aircraft. But Nimitz had reinforced the island, and aircraft based there have struck back at the enemy fleet, reporting multiple hits on enemy ships, including the aircraft carriers. You all know that such claims are often greatly exaggerated but, nonetheless, the possibility exists that the balance of air power has now tilted towards the Americans.

On the morning of 5 June Japanese cruisers are reported to be

UNITED STATES PACIFIC FLEET
USS PENNSYLVANIA,
FLAGSHIP OF THE COMMANDER-IN-CHIEF

June 3, 1942

S E C R E T

From: Commander-in-Chief,
United States Pacific Fleet

To: Commander Striking Force
(Operation Plan 29-42)

Subject: Operational Directive

Advance with all due haste to position northeast
of Midway and seek best possible tactical
advantage over the enemy. You will be governed
by the principle of calculated risk, which you
shall interpret to mean the avoidance of
exposure of your force to attack by superior
enemy forces without good prospect of inflicting,
as a result of such exposure, greater damage to
the enemy.

C. W. Nimitz

Source 3: Nimitz's Calculated Risk directive, May 1942

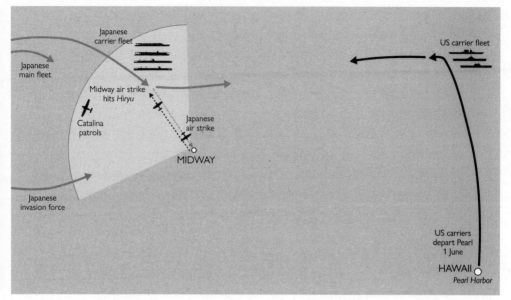

Map 5: US Navy's response, 4 June 1942

bombarding Midway, presumably preparatory to an amphibious assault. Shortly afterwards Catalinas once again begin sending reports about the position of the Japanese carrier fleet to the north of Midway; it seems there are now only three enemy carriers operating. Maybe one has indeed been damaged, or even sunk, by the Midway air strikes?

Admiral Fletcher orders an all-out attack on the Japanese carriers, and over the next hour a multitude of Dauntless dive bombers, Devastator torpedo bombers and Wildcat fighters launch from the American carriers. You are fully trained in air operations and watch appreciatively as they head out west to attack the enemy. Shortly afterwards, however, one of the combat air patrol Wildcat fighters providing protecting cover for the American fleet reports the presence of a lurking Japanese reconnaissance aircraft – the enemy may well now know your location. You begin ordering preparations for the *Yorktown* to ready itself to defend against an air attack in the next few hours.

Radio silence is generally being maintained to avoid giving anything away to the enemy, so you are reliant on reports from your air-group commanders as they press home their attacks against the enemy fleet. But much like the Coral Sea battle a few weeks earlier, there is a great deal of confusion and exaggeration. Some of your pilots signal that heavy damage has been inflicted on the enemy, and later that day Catalina flying-boat scouts report that at least one Japanese aircraft carrier – the *Kaga* – has been left ablaze, while signals intercepts tell you that another, the *Hiryu*, has already been damaged and forced to withdraw.

But the counter-strike soon falls. Despite the elation at having apparently caused some significant damage to the enemy, your radar operators start sounding the alarm about incoming Japanese strike aircraft, almost certainly from the enemy's carriers – it seems as though your air strike did not prevent the enemy launching simultaneously at you. Despite the best efforts of your defending Wildcat fighters, very quickly a maelstrom of anti-aircraft fire, diving and weaving aircraft and frantically manoeuvring ships erupts around and over the US fleet. You watch from the bridge of the *Yorktown* as the *Hornet* is targeted by enemy dive bombers and disappears in a deluge of spray and smoke – she appears to be badly damaged. Then your own ship is targeted, and Japanese torpedo bombers carry out a skilfully executed attack from two directions. Despite your orders to carry out emergency manoeuvres, the *Yorktown* is hit twice, followed by more hits from enemy dive bombers. Patched up as she is, the *Yorktown* is unable to sustain such damage easily and fires erupt across the vessel, forcing you to give the order to abandon ship.

Admiral Fletcher quickly transfers his command to a nearby cruiser and, as per tradition, you – as the captain – are last to leave the stricken aircraft carrier. You use scrambling ropes to reach the lifeboats, a task made much more difficult by the ship's alarming list to port. The survivors of the *Yorktown* are recovered by nearby cruisers and destroyers, but the aircraft carrier stays stubbornly afloat for many hours, prompting some to wonder whether your order to abandon ship was a little premature. You and your staff begin to consider whether the ship might be saved, but matters are taken out of your hands when a lurking Japanese submarine fires two more torpedoes into the stricken carrier on 6 June and the *Yorktown* finally slips beneath the waves shortly afterwards.

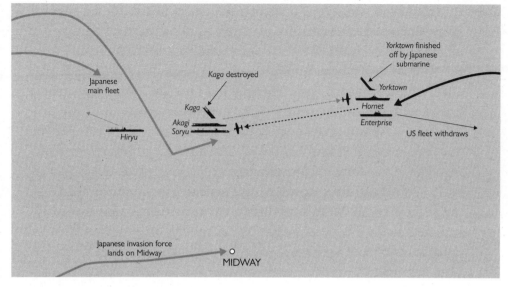

Map 6: Outcome, 5 June 1942

AFTERMATH

With only one functioning aircraft carrier remaining, Admiral Nimitz was left with no option but to withdraw his fleet. The Japanese eventually landed on Midway, but the main result the Japanese had been seeking – the destruction of the entire US carrier fleet – had not been achieved.

The Battle of Midway had been a damaging but frustrating draw for both sides. Nimitz was annoyed: a glorious opportunity to ambush the Japanese fleet had been frittered away and although Midway was hardly a crushing defeat, it had proved to be no more than an attritional exchange. Such an outcome favoured the USA in the long term, as they had the greater industrial production and training capacity to replace the losses, but the campaign in the south-west Pacific for Guadalcanal that erupted in the autumn of 1942 was made that much more demanding because of the indecisive outcome at Midway.

Yet for the Japanese, Midway hardly proved to be the ringing endorsement of the Imperial Japanese Navy's strategy – one big fleet aircraft carrier was lost and another put out of action for well over a year. Midway was eventually captured by the Japanese, though not without a bitter fight and heavy losses, but the longer-term gains were not so obvious. When Admiral Isoroku Yamamoto, Commander-in-Chief of the Japanese fleet, whose grand plan the Midway operation had been, reviewed the results of his endeavour it was all a little disappointing. An important outpost from which to oversee Pearl Harbor had been seized, but it would prove to be a great drain on resources to maintain it, as some of Yamamoto's critics had predicted; Midway was to be repeatedly suppressed and attacked by American forces in the years to come. More importantly, the need to neutralise the American fleet in the Pacific, to secure Japan enough time to strengthen its defences and build up its resources for the longer war ahead, had not been achieved.

THE DECISION

▶ To follow the alternative route, go back to **Section 2** (p. 184).

OR

▶ To explore the history of the Battle of Midway, go to the **Historical Note** on p. 221.

SECTION 4
SPRUANCE'S HESITATION

0750 (GMT −12 hours), 4 June 1942, USS *Enterprise*

There is lingering tension and frustration on the bridge of the USS *Enterprise*. You, **Captain Miles Browning**, have clashed badly with Rear Admiral Ray Spruance, who is leading Task Force 16. Despite your best efforts to persuade him to start launching your air strike against the Japanese fleet around 0715, Spruance held back, wanting to take the safer option of launching later, closer to 0900. The risk of launching early was too great and could result in the loss of many valuable aircraft and air crew and sacrifice any realistic option for follow-up actions, he announced. Not someone to hold back, you openly object, but the quietly spoken Spruance is clear and calm and points to Nimitz's guidance about taking calculated risks. You and your staff begin planning to launch around 0830, which in your view is much too late to maximise effectiveness, but Spruance is in charge.

Matters take another turn, however, at 0740 when your covering combat air patrol of Wildcat fighters flying over the US fleet identifies a Japanese reconnaissance aircraft lurking over Task Force 16; there is now every likelihood that the enemy knows your position and will be planning to attack with all haste. You have to launch immediately, in your view; further delays could spell disaster. Spruance now agrees. Inwardly you fume at the delay imposed by Spruance's 'hesitation', as you see it. Your air strike will now begin assembling a little after 0800 – an unnecessary delay of some forty-five minutes. Hopefully it won't prove decisive.

Hornet sends up thirty-five Dauntless dive bombers and sixteen Devastator torpedo bombers, escorted by eight Wildcat fighters, all led by Commander Stan Ring; meanwhile the *Enterprise* launches thirty-three Dauntlesses, fifteen Devastators and six Wildcats, with Lieutenant Commander Wade McClusky leading the force. It takes until 0840 to get the air groups into the air and properly coordinated, by which time further reconnaissance reports have fixed the position of the Japanese carriers more accurately. However, you are a little concerned that the launch has been rushed and that McClusky and his combined air group appear to be somewhat disorganised. In addition there has been little coordination with *Hornet*'s air group. Better news comes when Rear Admiral Fletcher on *Yorktown* signals that he has recovered his scout aircraft and is also in a position to launch an attack at the Japanese. By a little after 0900 thirty-eight more aircraft are winging their way towards the enemy fleet.

The senior staff on *Enterprise* wait anxiously for news and some indication of success. Radio silence is largely maintained while at sea, so you hear very little until your air-group commanders and pilots start sending back sketchy reports around 1045. McClusky signals back that two large enemy aircraft carriers have been hit multiple times and left on fire. Ring, commanding the *Hornet*'s air group, is suspiciously quiet, but the *Yorktown*'s Dauntless group, led by Lieutenant Commander Max Leslie, also signals that carriers are on fire and that other ships have been damaged.

Elation breaks out among Spruance's staff, but you, along with other officers well versed in carrier air-group tactics, are conscious that the Japanese may well have had enough time to launch their own air strike before yours hit home – you are not out of the woods yet. A little after 1130 your radar teams start to issue reports that there are multiple incoming contacts, undoubtedly the enemy. The American fleet prepares itself to be attacked – safety measures are put in place, anti-aircraft guns are readied and more Wildcat fighters are launched to supplement the existing combat air patrols. The ships move into defensive groupings to maximise the effects of anti-aircraft fire. You scan the skies for the incoming air strikes.

Despite the best efforts of your defending fighters and anti-aircraft

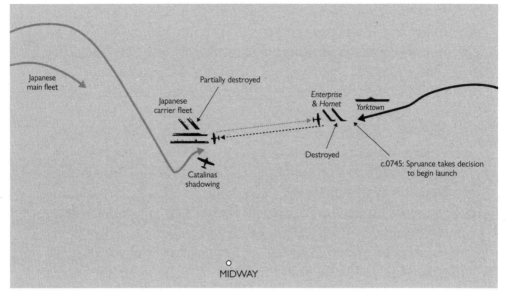

Map 7: Outcome, 4 June 1942

gunners, a proportion of the Japanese aircraft break through and begin launching well-coordinated torpedo and dive-bombing attacks on the *Hornet* and *Enterprise*. Frantic manoeuvres evade some of the attacks, but the *Enterprise* shudders when hit by a torpedo, while two bombs plunge into the flight deck. Fires and explosions erupt and the ship begins to list alarmingly to starboard. Admiral Spruance is forced to transfer his flag, along with you and the rest of his senior staff, to a nearby cruiser and shortly afterwards the order is given to abandon the *Enterprise*. Later that day, as you look on, the carrier is torn apart by further internal explosions and the following morning she finally slips beneath the waves of the Pacific.

AFTERMATH

Both the *Enterprise* and the *Hornet* were lost in the Japanese attack, but the Japanese fleet had also lost two aircraft carriers: the *Akagi* and *Soryu*. It was later confirmed that the Japanese carrier commander, Admiral Chūichi Nagumo, had dithered on the morning of 4 June, just like Spruance. He held back his aircraft until a properly organised and coordinated attack could be mounted, as per the Japanese navy's set doctrine. This delay enabled the Americans to get in the first blow, but Nagumo was afforded just enough time to launch his air strike before the blow fell – a window created by Spruance's decision to delay his launch until he reached a safer distance.

As the American fleet withdrew, covered by the surviving *Yorktown*, Nimitz regarded the battle as both a missed opportunity to deliver a decisive blow and a significant blunting of Japan's offensive capability. The enemy had lost two of their big carriers, which they would find much harder to replace than the Americans would, when replacing their losses. The Americans' massive naval building programme promised large numbers of replacement carriers in the coming two years, to a level that would dwarf the efforts of the Japanese. In the short term, too, the Americans were able to boost the *Yorktown* with the returning *Saratoga* and the newly arrived *Wasp*. By the autumn the US fleet was operating in the south-west Pacific and the attritional battles around Guadalcanal would rumble on well into 1943.

For Admiral Ray Spruance, however, his moment was gone. He was soon replaced by Halsey and would forever be known for his hesitation at Midway – hesitation that caused the loss of two American carriers and enabled a surprised and exposed Japanese fleet commanded by a dithering admiral to evade total calamity.

THE DECISION

▶ To follow the alternative route, go back to
Section 2 (p. 184).

OR

▶ To explore the history of the Battle of Midway, go to the
Historical Note on p. 221.

SECTION 5
MCCLUSKY'S INTUITION

0920 (GMT -12 hours), 4 June 1942,
SBD Dauntless Group Commander,
c.150 miles north-west of Midway

In the skies over the Pacific Ocean a group of thirty-two Douglas Dauntless dive bombers of the US Navy are flying west at an altitude of some 19,000 feet. The pilots and gunners of each two-crew aircraft are scouring the sea below, vainly searching for the Japanese enemy's aircraft-carrier fleet, which they were expecting to see arrayed before them. You are *Lieutenant Commander Wade McClusky*, the forty-year-old leader of the USS *Enterprise*'s air group, and as the realisation dawns that the enemy are not below, you begin to wonder what might have gone wrong and, more importantly, what your next move might be. A critical decision is looming, and on it may depend the outcome of the whole Battle of Midway.

You have been a naval aviator since 1929 and have more than 2,900 flying hours under your belt. Although you have

7. Lieutenant Commander Wade McClusky

only recently taken up the role of leading the *Enterprise*'s air group, you have seen extensive action over the first few months of the war and have earned a string of medals and commendations. In fact just a week ago you had received a Navy Cross from Admiral Nimitz himself on the flight deck of the *Enterprise*. Modest and uncomfortable with self-promotion, you rely on quiet competence to command respect and exert authority. You are leading *Enterprise*'s air group from a Dauntless dive bomber, and although you are more experienced in fighter tactics, you also have considerable experience of dive-bombing. All this experience is about to be tested. This morning you have flown west-south-west from the carrier USS *Enterprise* to the estimated point of contact with the Japanese carriers that had been located earlier that day – but now all that you can see is empty ocean. There is no sign of the Japanese fleet. Could the gamble, or the calculated risk, that Admiral Spruance had made just three hours ago by launching his air attack at extreme range be about to backfire? The whole outcome of the battle is hanging in the balance, and you fully realise the enormity of the situation unfolding.

You had been informed of the operational and intelligence situation at a conference with Captain Miles Browning on the morning of

8. Douglas Dauntless

29 May. He briefed you and your fellow air-group commanders about expectations and tactical situations that were likely to emerge over the next few days, and that the morning of 4 June would in all probability be pivotal. One of your squadron commanders, Lieutenant Commander Dick Best, expressed scepticism about the detailed intelligence, a view that drew short shrift from the intimidating Browning. They did not get on.

On this morning of 4 June you were woken around 0200, to a breakfast of eggs and steak, although some of the nervous pilots merely pushed their food about their plates. By 0400 all the air crews were in their smoke-filled ready rooms, awaiting updates and orders. The pilots had plotted the *Enterprise*'s position, but the mood was pensive and a little subdued. You felt some irritation when it emerged that your usual gunner / radio operator, John O'Brien, had lost his glasses and will not be able to fly with you; instead you would be accompanied by a rookie, Walter Chocalousek, who only joined the squadron a few days earlier. The teletype machines then sprang into life a little after 0430, conveying the first sightings of the Japanese fleet. Over the next ninety minutes the pilots waited tensely for the order to get airborne; you noted some irritation when a premature call to arms at 0550 was quickly cancelled, exasperating all concerned. Eventually, however, as more accurate sighting reports filtered through, a little after 0615 you were called to briefings for all the pilots (though not the gunners), where your missions were outlined. Admiral Spruance was planning to launch a major strike as soon as possible, and the Dauntless dive bombers of *Enterprise*'s air group would be the first to begin getting airborne, from a little after 0700. Browning informed his commanders that there was every possibility their air group would hit the Japanese fleet in the midst of them recovering their air strike from Midway, thus making them hugely vulnerable. But an early launch was going to create further problems for the American air groups as they started to muster – problems that threatened to lower the chances of the attack succeeding.

Naval air-attack tactics in 1942 centred on two main methods of attacking and sinking enemy ships at sea. The most accurate method was dive-bombing, wherein single-engine bombers would arrive over an enemy fleet and zoom down from high altitude at a steep angle, then drop bombs at the enemy ship before swooping away, just above the waves, to escape. However, the most decisive way was with torpedoes launched at low altitude, by torpedo bombers almost hugging the water. Torpedoes would damage the ship below the waterline and were more likely to cause the ship to sink, whereas bombs, although more accurate, might cripple a ship, but were less likely to sink it. To complicate matters, torpedo-bombing was distinctly more dangerous, particularly when flying the American Douglas Devastator, which was slow and cumbersome. The Douglas Dauntless dive bomber, in contrast, was a decent and rugged aeroplane, much liked by its pilots and air crew. The key to success was to coordinate dive-bombing and torpedo-bombing attacks so that they hit at the same time, thus splitting the enemy's defences. To help the attack, the bombers should also be escorted by supporting fighter aircraft, whose role it was to try and keep enemy interceptor aircraft away from the bombers as much as possible. It was unlikely they would be able to keep the enemy entirely at bay, but everything had to be done to support the bombers.

The challenge that had confronted Spruance's air groups as they started to launch from the *Enterprise* and *Hornet*'s flight decks a little after 0700 was that the Devastator torpedo bombers and Wildcat fighters would be at the limits of their operating range. Therefore the Dauntless dive bombers would launch first, as they had the greater endurance, after which they would circle, waiting for the Wildcats and then the Devastators to get into the air, before they would all form up to head off and deliver their attack. Each carrier's air group of fighters, dive bombers and torpedo bombers was commanded by a small section of three aircraft headed by the air-group commander; in the case of the *Hornet*'s group, that was Commander Stan Ring, technically your superior, while you command the *Enterprise*'s combined force. You are, however, a little concerned that nothing has been arranged to coordinate your efforts with those of the *Hornet*'s air group, let alone that of the *Yorktown*'s, which is still some way off launching its aircraft.

Halfway through the launch, at 0745, with your pilots and air crew becoming a little frustrated as you wasted valuable fuel, circling the fleet waiting for the Devastators and Wildcats to get airborne, you received a message from Admiral Spruance: 'Proceed on mission assigned.' Something must have happened to cause you to be ordered to head off immediately, without waiting for the rest of the air group to form up with you. Perhaps an enemy scouting plane had been spotted lurking around the US fleet?

As you headed off, it was clear that the chances of organising a coordinated attack with your own air group had receded, to say nothing of *Hornet*'s force, which was now nowhere to be seen. Everything depended on the last sighting of the Japanese carriers being accurate and the assessments of where they might move to (they were unlikely to remain static) proving prescient enough to enable your air group to arrive in the right place at the right time. You would get no further updates, as you and the fleet would be keeping strict radio silence during your flight to the target to prevent the enemy detecting your approach.

Now, at around 0920, and after having been in the air for some two and a half hours already, your force has arrived at the point where they expected the Japanese fleet to be. They are ready to go into action, but agonisingly the Japanese fleet is nowhere to be found, nor is there any sign of Commander Ring and the *Hornet*'s air group, either. To add to your concerns, you have had no contact with the *Enterprise*'s torpedo bombers or fighters. You are on your own. Fuel is running dangerously low and time is running out. You check your navigation chart to make sure you are where you are supposed to be – it seems in order. Another check with your binoculars of the oceans around you: still nothing. You sense the growing concern of the other pilots and squadron commanders. There is no radio chatter but hand signals, shrugs and looks of concern abound.

You consider your options quickly. Either the Japanese fleet has pressed on south towards Midway, perhaps to close the range in preparation for a second strike, or it has turned away to the north to buy time or perhaps to follow up a sighting. You have enough flying time and fuel to commit to one of those options before having to head back to *Enterprise*. If you choose correctly, you will probably locate the Japanese fleet; if you get it wrong, you will have to return to

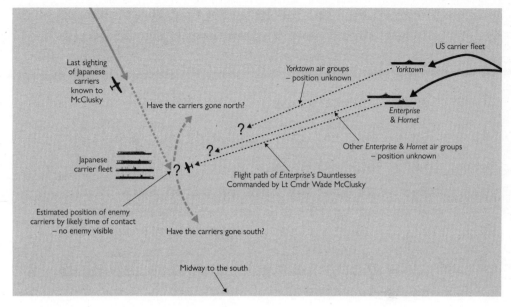

Map 8: McClusky's choice, 4 June 1942

Enterprise empty-handed. Indeed, failure might hand the initiative to the Japanese, who might then locate and destroy the *Enterprise* before you get back.

AIDE-MEMOIRE

* Consider the tactical situation (see Map 8).

* You only have fuel to make one choice – there is no second chance.

* The other air groups from *Enterprise* and *Hornet* are nowhere to be seen.

* The Japanese fleet could well have steamed on southwards towards Midway, as this would shorten the time required to pick up the air strike the Japanese had launched against the island earlier in the morning.

* But it is also plausible that the Japanese fleet could have turned away from Midway to buy more time to beat off air attacks from the island.

* A turn north means the Japanese are pulling back from Midway. Is that likely? Surely they would continue south to press home their attack?

* But what if they have received news of a sighting of the US fleets? A turn north would enable them to turn their attention to the greater threat of the American carriers, or even just gain some breathing space to regroup.

THE DECISION

Do you dip your wings to the right and head north away from Midway or, conversely, turn left towards the island? On your decision hangs the outcome of the Battle of Midway and possibly the whole war against Japan.

▶ If you decide to head north, go to **Section 6** (p. 212).

OR

▶ If you decide to head south, go to **Section 7** (p. 218).

SECTION 6
MOMENT OF TRUTH

0930 (GMT -12 hours), 4 June 1942, SBD Dauntless Group Commander, c.150 miles north-west of Midway

After a cursory scout around, you, **Lieutenant Wade McClusky**, decide to lead the *Enterprise*'s Dauntless dive bombers north, estimating that the Japanese fleet would have turned away from Midway for a while, possibly to regroup and to recover and re-equip its aircraft.

Around 0955 your intuition is rewarded with the sight of a single Japanese destroyer steaming north at high speed – it could well point the way to the Japanese fleet, you surmise. You also receive a radio message from Lieutenant Commander Jim Gray, commanding the *Enterprise*'s Wildcat escort fighters, which are circling the Japanese fleet ahead of your force. They have already located the enemy fleet in conjunction with the *Enterprise*'s Devastator torpedo bombers, although they had since lost contact with them. With a great sense of relief, you now realise that you have chosen correctly and very soon you sight the Japanese carrier fleet, looking rather ragged and disorderly, but nevertheless intact. There is no obvious sign of Japanese fighters, so you decide to press home your attack with alacrity on the two biggest carriers. At 1022 you push the nose of your Dauntless into a seventy-degree dive and begin lining up the huge flight deck of the Japanese carrier *Kaga* in your sights; behind you the rest of the air group also begin their dives, splitting their attacks between *Kaga* and the Japanese flagship, *Akagi*. Anti-aircraft gunfire from the ships does little to deter the attacks and the Dauntlesses release their bombs at around 2,500 feet before swooping away low over the sea, in some cases pursued by some now-alerted and frantic enemy Zero fighters. It is too late. The two principal aircraft carriers in the Japanese navy have been reduced to blazing wrecks.

After your diving attack your Dauntless skims the sea initially at around twenty feet, before heading away from the enemy fleet, southwards at first, as per the pre-take-off briefing. This tactic, it is hoped, will confuse the enemy as to where your aircraft has come from. However, you are still in the battle zone, and although fifteen minutes have passed and you are now flying east back to the *Enterprise,* you still come under attack from two Japanese Zero fighters. Although your gunner, Walter Chocalousek, managed to shoot one of them down – no mean feat – and the other eventually flew off, your Dauntless is hit more than fifty times and you suffer wounds to the shoulder.

As your Dauntlesses wing their way back to the *Enterprise* you

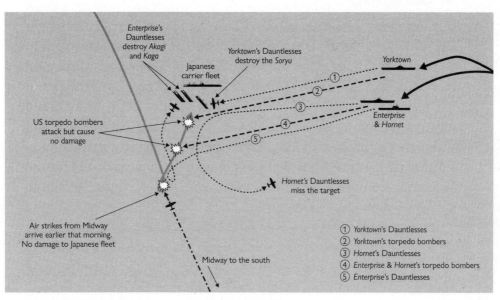

Map 9: Moment of truth, morning of 4 June 1942

realise that you have achieved a stunning success. You eventually arrive back at the US fleet and insist that, despite your wounds and damage to your aircraft, all the other surviving Dauntlesses from your group land before you. Eventually, at 1142, you arrive back on the *Enterprise*, having been in the air for more than four and a half hours – the absolute limits of the endurance of a Dauntless. After reporting back to Admiral Spruance, Captain Miles Browning and the rest of the senior staff, you are whisked off to the sickbay for treatment. Later you hear that the *Yorktown*'s dive bombers had also delivered a decisive attack against a third Japanese carrier, the *Soryu*, almost simultaneously with your own, leaving the Japanese fleet just one functioning carrier to the Americans' three. Nonetheless, American aircraft losses have been heavy. The Devastator torpedo bombers have suffered most heavily, and even your own group of Dauntlesses lost sixteen of their number to enemy fighters, anti-aircraft gunnery and lack of fuel, which caused them to ditch in the ocean.

AFTERMATH

The surviving Japanese carrier *Hiryu* launched two successful air attacks against the *Yorktown* on the afternoon of 4 June, causing it to be abandoned. It was later finished off by a lurking Japanese submarine. Counter-strikes from *Hornet* and *Enterprise* destroyed the *Hiryu*, however, which was itself later abandoned and scuttled by Japanese torpedoes. The Japanese abandoned their assault on Midway and withdrew, having lost all four of their frontline aircraft carriers. It was a disaster for them, although it was not until the autumn of 1942 in the Solomons campaign around Guadalcanal that Japanese offensive strength was entirely extinguished.

How had McClusky's – and indeed the *Yorktown*'s – dive bombers achieved such success at Midway? In part it was a result of the *Enterprise* and *Yorktown* dive bombers arriving over the Japanese fleet almost simultaneously, despite the *Yorktown* Dauntlesses having launched an hour later than the *Enterprise* group. The *Yorktown*'s air group headed out on a more direct northerly bearing and so, when the

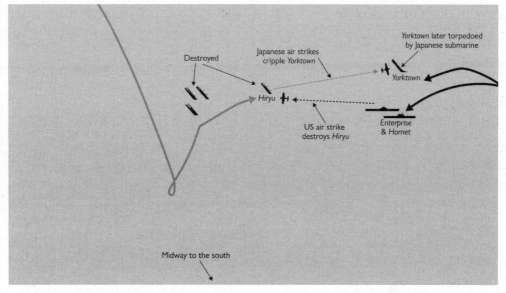

Map 10: Final moves, afternoon of 4 June 1942

Japanese fleet turned north away from Midway, it had located them much sooner.

Both groups of dive bombers met little resistance during their attacks and had thus delivered devastating blows to the Japanese fleet, largely unmolested, but how had this happened? The path was opened up for the Dauntlesses by the actions of the air groups from Midway, which had attacked the Japanese fleet relentlessly that morning, at very heavy cost and with no discernible success. They had then been followed by the Devastator torpedo bombers from the three US carriers, which had pressed home their attacks, mostly without fighter-escort cover, and had suffered calamitous casualties – one squadron suffered the loss of all fifteen of its Devastators, with only one man surviving. Their combined efforts eventually bore fruit, as all these attacks had drawn many of the defending Zero fighters down to low altitude, leaving the skies high above open for the Dauntlesses to deliver their battle-winning strikes. It was a significant misjudgement on the part of the Japanese fighter directors to allow this to happen, but the Japanese were also taken by surprise in part because their fleets lacked early-warning radar.

But the Japanese leadership had also blundered badly during the battle. Firm in their belief that no US fleet was at sea, Japanese reconnaissance had been a little cursory: one aircraft took off late; another missed the US fleet, despite flying over it; and the incoherent initial reports from the aircraft that eventually did sight part of the US fleet merely sowed confusion. Admiral Chūichi Nagumo, commanding the Japanese carriers, had also dithered at crucial moments, causing critical delays in the Japanese response to the US attacks. He was, in truth, sticking rigidly to Japanese carrier doctrine, but he showed a lack of flexibility. He almost got away with it, however; his strike force was close to being ready to launch when the American dive bombers hit. But his caution and delay proved fatal.

Yet the Japanese were also fortunate, for what had happened to *Hornet*'s Dauntless dive bombers led by Commander Stan Ring? They took no part in the morning's actions against the Japanese fleet. Ring and his Dauntlesses had in fact set off on an erroneous heading, one that the *Hornet*'s torpedo bombers and fighters contested and then ignored. Consequently Ring missed the Japanese fleet entirely, made the wrong choices and ended up leading his straggling group home to the US carriers having achieved nothing. It seems possible that various officers then covered up the miscalculation for many years.

McClusky also came in for some scrutiny. He may well have acted too impulsively when he began the attack that morning, as he targeted the nearest enemy carrier, whereas doctrine should have caused him to lead the first group of dive bombers to the further-away target, to enable his following dive bombers to target the nearer ship. It did not help that McClusky's bomb missed the target. Nonetheless, as Admiral Nimitz later stated, McClusky's intuition to head north that morning when the Japanese fleet was not where he had expected 'decided the fate of our carrier task force and our forces at Midway'. It was a decision that determined the outcome of the Battle of Midway.

THE DECISION

▶ To follow the alternative route go to **Section 7** (p. 218).

OR

▶ To explore the history of the Battle of Midway go to the **Historical Note** on p. 221.

SECTION 7
MISSED OPPORTUNITIES

0935 (GMT -12 hours), 4 June 1942, SBD Dauntless GC, c.150 miles north-west of Midway

After scouting around without luck for the Japanese fleet, at 0935 you, *Lieutenant Commander Wade McClusky*, eventually decide to turn your air group south-south-east towards Midway, reckoning that the enemy must have attempted to close the distance to its objective. It still concerns you that you have had no indication about the progress of the other air groups from *Enterprise* or *Hornet*. As you head south, you hope and expect to see the Japanese fleet appear ahead of you but, as the minutes roll by, you start to doubt your choice; there is no sign of the enemy. You are getting too close to Midway Island itself by now – clearly you have blown your chance. In despair you turn back to the American carriers, hoping that you have enough fuel to make it. En route you encounter some straggling Dauntlesses from the *Hornet*'s air group. Both strike forces appear to have missed their targets, and both are having to trail back to their carriers away to the north-east. Some aircraft choose to refuel on Midway, as they doubt they have the reserves to make it home.

With the Dauntlesses out of the picture, what of the Devastator torpedo bombers from the American carriers? And what of *Yorktown*'s air group? The Devastators from *Hornet*, *Enterprise* and *Yorktown* quickly and intuitively chose the correct route to the Japanese fleet, but were soon confronted by stout and aggressive defences, suffered heavy losses and failed to secure any significant hits. These sacrifices

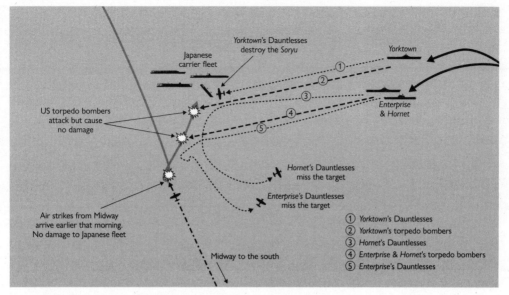

Map 11: McClusky turns south, morning of 4 June 1942

are not in vain, however, as, with all the Japanese fighters down at low level, battling against the American torpedo bombers, the skies over-head are completely open. The *Yorktown*'s newly arrived Dauntless dive bombers sweep down from 14,000 feet and drop a pattern of bombs that devastates the carrier *Soryu* in a matter of seconds. The crippled ship is soon torn asunder by massive internal explosions and sinks later that day. But by then much has changed.

AFTERMATH

The Americans had achieved some success, but the failure of McClusky's and Ring's groups to find the enemy fleet resulted in a tremendous missed opportunity; the Japanese had been hit, but they were not out. Smarting from the loss of the *Soryu*, the Japanese fleet struck back with a vengeance. They now had accurate sightings and reports on the US fleet, and they launched a strike of some 140 air-craft against Spruance and Fletcher's carriers. They hit later that morning and, despite a desperate and frantic defence, *Yorktown* was

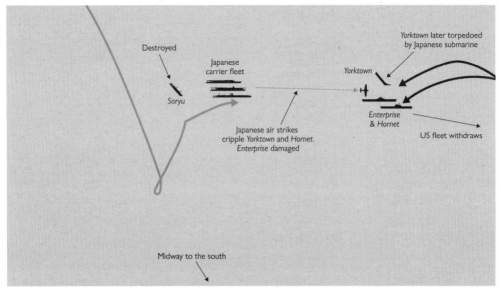

Map 12: Outcome, afternoon 4 June 1942

severely damaged and left ablaze, while *Hornet* was crippled by three torpedoes and soon began to list alarmingly. *Enterprise* survived with minor damage, although Nimitz had no option but to withdraw what remained of his fleet. *Yorktown* was eventually finished off by a Japanese submarine, and while *Hornet* limped back to Pearl Harbor, she was out of action for many months.

Yamamoto's force finally captured Midway after a brutal and highly destructive amphibious assault, but although he had accounted for one American carrier, this was not the decisive engagement he had hoped for. Two enemy carriers had escaped, and the Japanese themselves had lost a valuable asset in the *Soryu*. In part this was down to his trap having being sprung by the Americans, but it could have been much worse; if McClusky and Ring had made different choices that morning, the outcome of the battle would have been very much in the Americans' favour.

In strategic terms, despite Yamamoto's great gamble, little had in fact changed. Midway proved to be a strain on Japanese resources and was constantly harried by US forces operating out of Pearl Harbor, while as the summer progressed, the Americans added *Saratoga* and

Wasp to their carrier force. However, they were forced to postpone any offensive actions in the South Pacific that had been planned for 1942, but by the following year they had landed at Guadalcanal and a series of attritional naval battles had begun that would slowly whittle away Japan's strength. By 1944 the Americans were fully on the offensive. Ultimately, Midway had decided nothing.

THE DECISION

▶ To follow the alternative route, go back to **Section 5** (p. 204).

OR

▶ To explore the history of the Battle of Midway, see the **Historical Note** below.

HISTORICAL NOTE

On 4 June 1942 the US Navy won a tremendous victory against the elite Japanese carrier fleet at Midway: all four Japanese carriers were destroyed for the loss of just one American carrier, the *Yorktown*. The Japanese abandoned the assault on Midway itself and withdrew. How had this come about? The historical path to victory was to embrace the intelligence of Layton and Rochefort and place the US carriers north-east of Midway, lying in wait for the unsuspecting Japanese fleet on 4 June (from Section 1, go to Section 2). From there you had to take the gamble of an early launch against the Japanese fleet – it was essential to hit them as quickly as possible *before* they had a chance to locate you and initiate their own attack, whatever the risk to your own air group (from Section 2, go to Section 5). Then in your Dauntless dive bomber, as Wade McClusky, you had to choose north rather than south (Section 6, not Section 7) to achieve the decisive result.

If you chose alternative paths, such choices were not without merit and far from foolish. Nimitz took a considerable gamble placing his carriers so close to Midway – if the intelligence was flawed or misleading, the Japanese might have delivered a fatal blow to Hawaii. If the US carriers had been too far away to intervene, the outcome could have been damaging for the USA. Another alternative worth pondering is: what if the Japanese had realised the Americans had deduced that Midway was the target? What if the Japanese had become suspicious about the fresh-water ruse; or noticed that the Americans had increased radio traffic, indicative of fleet activity, or peculiarly long air reconnaissance patrols operating out of Midway in the period leading up to the battle? If the Japanese fleet had been alerted and had been lying in wait for the US carriers, the outcome could well have been the decisive result that Admiral Yamamoto, commanding the Japanese navy, had been hoping for. However, Japanese intelligence-gathering, filtering and processing was deeply flawed, so such an outcome was highly unlikely, but it could not be ruled out – and some in Washington made such warnings.

What of Spruance delaying his launch on the morning of the battle (from Section 3, go to Section 4)? Despite the likely explosion from the irascible and outspoken Browning, such a choice was not without consideration. Spruance had one chance to get in a terrific blow against the enemy – get it wrong and the battle would be lost. It was a major gamble to launch early and although it proved decisive, there was still the possibility that it might have gone dreadfully wrong, which would have left the American carriers hopelessly exposed. Additionally, at the time of launching it was unclear how many enemy carriers had been located.

Finally, what if McClusky had turned south, unwittingly away from the Japanese fleet (from Section 5, go to Section 7)? Such an outcome might well have resulted in the *Soryu* still being sunk by the *Yorktown*'s Dauntlesses, but the other three carriers would have escaped, two of which in the historical path were destroyed by McClusky's air group. With three Japanese carriers still in play, the Americans would have been vulnerable to a major air attack by superior elite forces, and the chances were that two – or even all – of the US carriers might have been destroyed in the late morning / early afternoon of 4 June. Conversely, what if Commander Ring's *Hornet* Dauntlesses had not got 'lost' and had located and attacked the Japanese fleet at the same time as McClusky? With another thirty-plus Dauntlesses attacking, there was every chance that all four Japanese carriers would have been knocked out in one fell swoop.

The alternative choices offered here largely result in indecisive outcomes, which follow the pattern of the Coral Sea battle and were later mirrored in the naval actions around the Solomon Islands in the south-west Pacific in the late summer and autumn of 1942. Midway was different in that American intelligence provided the US fleet with an opportunity to secure a decisive advantage, but it was no more than that: an opportunity. It still required clear, incisive and intuitive thinking from Spruance and his staff, and from McClusky.

The Battle of Midway is often cited as the crucial turning point in the war that raged between Japan and the USA across the Pacific between 1941 and 1945 – perhaps overly so. The long-term foundations of America's ultimate victory were in place even before the war broke out, and even if Midway had ended in defeat, its ultimate success was still likely. What Midway certainly achieved, however, was the blunting of Japan's offensive power. The Americans continued to press the war in the Solomons in the autumn of 1942, but this soon slipped into a grinding attritional campaign that was always going to favour the much stronger industrial and organisational capacity of the USA. Midway is also often described as a 'miracle', with a small, determined US force outfighting a much larger and better-equipped Japanese leviathan. In truth, in the critical category of aviation, the two sides were quite evenly matched, and the battle was to be decided by the correct and calculated application of air power. It was in this intelligence-informed calculation process that the Americans triumphed.

Captain Lynde D. McCormick enjoyed a successful naval career, despite fracturing a vertebra just a few weeks after Midway. He became a Rear Admiral in July 1942 and won two Legion of Merit awards. He commanded capital ships in action in the Atlantic and the Pacific, and enjoyed a successful post-war career until his untimely death at sixty-one in 1956.

Captain Miles Browning was highly unlikely not to have pressed Spruance to attack all out on the morning of 4 June, and in this he was spot-on in his estimation. However, his judgement in other aspects of his personal and professional lives was not always so smart. Shortly after Midway it emerged that he had conducted an affair with a fellow officer's wife (who later became his fourth and final wife), and his ongoing heavy drinking and mercurial nature blunted his progress. He remained deeply unpopular with many of those who served with and for him.

Captain Elliott Buckmaster was promoted to Rear Admiral soon after Midway and went on to a successful career leading naval air-training programmes. He was later involved in the rescue operations in the wake of the *Indianapolis* tragedy in 1945.

Lieutenant Commander Wade McClusky won a Navy Cross for his action at Midway, and later enjoyed a successful career throughout the rest of the Pacific War. He served in the post-1945 US Navy and was promoted on retirement in 1956 to Rear Admiral.

5

BOMBER OFFENSIVE:
STRATEGY, MORALITY AND THE ROAD TO DRESDEN, 1940–45

SECTION 1
THE BOMBER WILL ALWAYS GET THROUGH

3 July 1940, Defence Committee meeting, Cabinet War Rooms, Whitehall

Churchill's Defence Committee has assembled once more to discuss the pressing matters confronting Britain. Around the table sit the military Chiefs of Staff, the political heads of the three armed services, General Hastings Ismay (Churchill's military advisor) and you, *Lord Beaverbrook (Max Aitken)*. You are a Canadian press baron – you own the largest-circulation newspaper in the world, the *Daily Express* – and multimillionaire, who has been a leading figure in British and Commonwealth life and society since before the First World War. Having previously served as an MP, you were elevated to the Lords in 1917 and have served in a range of high-profile positions ever since. Because you are known as a shrewd businessman and first-class organiser, Churchill recently brought you into government as Minister of Aircraft Production, such is the importance of aviation to British strategic needs. Although you are a member of the War Cabinet, you do not usually attend the Defence

1. Lord Beaverbrook

Committee meetings, but Churchill has specifically asked for you on this occasion.

The mood of the room is sombre. The war is not going well – the British Expeditionary Force may have been saved, although it is now bereft of much of its equipment, but the United Kingdom's ground forces and defences are at breaking point. The Germans are clearly massing for an air and seaborne attack on the British Isles and everything now depends on the Royal Navy and the Royal Air Force keeping the Nazis at bay. The Chiefs of Staff had stated a few weeks earlier that they considered that even after the collapse of France, Britain could still survive. That is about to be put to the test.

Although the meeting is focused on the immediate survival crisis, there is nonetheless recognition that a means of hitting back at Germany, perhaps even a route to eventual victory, has to be identified. But with what? The British Army alone cannot take on the much larger and better-equipped Germans, and although the Royal Navy dominates the seas, it can only provide a supporting role in any path to victory. The Royal Air Force, however, offers an alternative and modern option for taking the war directly into Germany. Cyril Newall, the Chief of the Air Staff, presses the idea of releasing the RAF's Bomber Command to a more wide-ranging and full-throttle assault on German industry, possibly resulting in a direct attack on the enemy's morale. An expanded and determined bombing campaign against Germany itself would bypass the German army and allow Britain to hit back directly from the British Isles, with no immediate need to return to fighting on the continent.

It is not a new idea by any means. The Germans had attacked England in the First World War with airships and bombers, and the world's air forces had wrestled with the idea of independent bombing fleets directly attacking enemy industries and cities throughout the interwar years. Bombing had captured the attention of the world by the 1930s, fuelled by the likelihood of poison gas being deployed against civilians. In 1936 the film *Things to Come*, based on an H. G. Wells novel, had graphically and disturbingly portrayed the devastating impact of the bombing of civilians, the effects of poison gas and the probable disintegration of public order. Concern about bombing had been growing throughout the 1930s, and efforts to limit or ban the use of bombers had been a central part of disarmament talks. The British Prime Minister Stanley Baldwin had famously highlighted the threat in 1932: 'The bomber will always get through,' he claimed, as a device to press for progress in disarmament. In 1937 the infamous bombing of the small Basque town of Guernica, in northern Spain, during the Spanish Civil War had highlighted some of the horrors and consequences of bombing towns. As one writer put it, 'the shadow of the bomber' loomed large over the 1930s.

Yet by the outbreak of the Second World War the RAF in Britain, although enthusiastic in its support for long-range bombing, lacked suitable aircraft and technical know-how. Plans are in place to build more and better heavy bomber aircraft, a campaign you are now leading, but delivering success with the aircraft currently available seems a tall order. So far bombing operations against broadly military targets have been less than successful. The RAF had fewer than 300 bombers

2. Handley Page Hampden

in September 1939, too few in number to have a major impact, and the main types of bomber aircraft available – Handley Page Hampdens, Vickers Wellingtons, Armstrong Whitworth Whitleys and Bristol Blenheims – were relatively slow, under-gunned, short-ranged and could carry only a limited payload of bombs.

Equally important, however, is bombing policy. What should the RAF bombers actually be targeting? The apocalyptic visions of poison gas and indiscriminate bombing of civilians have largely been rejected as morally and politically unacceptable by the world's governments. Air-force staff themselves have pressed the notion of precision bombing of carefully chosen industrial and economic targets – those most likely to have a profound effect on the enemy's ability to wage war. Yet thus far in the war, even when allowed to try, such bombing raids have ended in dismal failure and heavy losses in RAF aircraft and personnel.

You have heard that doubts have emerged about daylight precision bombing in the RAF, particularly after heavy losses over Norway in April. The previous Commander-in-Chief of Bomber Command, Air Chief Marshal Edgar Ludlow-Hewitt, was an intelligent and thoughtful man, deeply sensitive to the losses and difficulties being sustained by his air crews. He had resisted committing to any determined campaign against the economic heart of Germany, the Ruhr, until such time as it might be managed with sustainable and tolerable losses. One way that losses might be kept under control would be to switch even more to night-time bombing operations; by operating in darkness, it is hoped that bomber losses to enemy fighters would be reduced. The downside to this shift in tactics, however, is that navigation and bombing accuracy will be reduced still further in darkness. Technology may in the long term offer some solutions to improving targeting accuracy, but currently visual navigation and bombing are the only real option – and bombing in darkness makes this much less successful. It is certain that bombing raids will become less accurate and therefore more likely to miss their primary targets and hit surrounding areas, probably killing civilians. Expanding and widening the bombing campaign will undoubtedly bring with it criticism over increased civilian casualties.

As Newall explains the issues to the Defence Committee, it is clear that the RAF's view is hardening. Ludlow-Hewitt had been replaced as Commander-in-Chief of Bomber Command by Air Marshal Charles Portal, partly because Portal is willing to commit the bombers to difficult and potentially risky operations. Portal – the 'accepted star' of the RAF – is a highly intelligent commander who studied at Oxford before joining the Royal Engineers in 1914. He flew some 900 missions during the Great War, and later became a key figure in the interwar RAF. You have heard that he is an enthusiastic supporter of ramping up operations against German industrial targets, particularly synthetic-oil production plants. He and Newall have been dismayed by the careful shackles placed on targeting policy – they are now arguing that if a sustained long-range bombing campaign is to provide a route to victory for Britain, the gloves need to come off and bombing policy needs to be relaxed to enable wider areas to be targeted. They accept that there would be an increase in civilian casualties, but the Germans have at times openly targeted cities for strategic effects – Rotterdam in May 1940 being a case in point – and the damage to German civilian morale from intensified British bombing raids is a factor to be considered useful. Attacking enemy morale had always been a possible option in air-war theory in the interwar years, after all.

Portal has, however, been told by the Air Staff that 'in no circumstances should night bombing be allowed to degenerate into mere indiscriminate action'. The British government is determined to draw upon the support of the wider world, most obviously the USA, and reckons that an obvious all-out bombing campaign against morale, in which the Germans would be able to point to British barbarity and cruelty against 'defenceless' civilians, would do enormous harm.

What is to be done therefore? As Churchill and his War Cabinet Defence Committee debate the wreckage of Britain's strategy, it is clear that an intensified and expanded bombing campaign against the German economy is a possible option. It will require a huge investment to make it work, and you, as Minister of Aircraft Production, make this clear.

AIDE-MEMOIRE

* There are huge tactical and technical
  difficulties to overcome in delivering the
  bombing campaign, difficulties that will
  require the investment of huge resources,
  at a time when Britain is almost on its
  knees.

* But what other better options does your
  government actually have?

* Is there a case for suspending the grand
  bombing campaign and focusing the RAF's
  efforts and resources on tactical support
  operations for the army and the navy?

* A sustained bombing campaign against the
  German economy might yield great strategic
  benefits, *if* it could be made to work.

* Is it shrewd to invest such a great deal
  in an as-yet-unproven strategy? It has
  never been tried, and therefore no one
  knows whether it will work.

* The benefits for morale at home if the RAF
  is seen to be hitting back against
  Hitler's Germany will be huge.

* It appears that the only way to make the
  bombing campaign succeed will also
  increase civilian loss of life and bring
  down the ire of world opinion, just as you
  need global support most.

* The limited night-time campaign against
  Germany's industries is proving no more
  successful than its daylight operations.

* You need the Americans to support you,
  after France's surrender on 22 June.
  Bombing shows that you are worth backing
  because you are still fighting; no one in
  Washington (or elsewhere) will offer
  support to a lame duck that does not seem
  to have a way of fighting back, let alone
  winning.

THE DECISION

What advice do you offer to the Defence Committee?
You are confident that you can provide aircraft for many roles
and strategies, but what do you think should be the
government's policy at this time? Ultimately is it time for the
gloves to come off and for Britain to push the boundaries of
modern warfare?

▶ If you want to press on with an expanded and intensified
  bombing offensive, go to **Section 2** (p. 234).

  OR

▶ If you want to switch your air-power resources into
  different areas, go to **Section 3** (p. 242).

SECTION 2
FAILING STRATEGIES

18 August 1941, Air Ministry, Berkeley Square, London

The Air Ministry, the political and administrative centre of the Royal Air Force, is a hive of activity, with civil servants and staff bustling about. But within the senior membership of the department there is considerable tension, bordering on dismay. The Air Ministry has just received a War Cabinet Secretariat report on the RAF's bombing offensive against Germany, a report initiated by Frederick Lindemann, the Prime Minister's mercurial chief scientific advisor and therefore one who has to be taken seriously. The report, soon known as the Butt Report after its author, the economist David Bensusan-Butt, one of Cherwell's assistants, does not make easy reading for anyone connected with the prosecution of Bomber Command's campaign (see Source 1 on p. 235).

You are **Arthur Street**, the forty-eight-year-old Permanent Secretary at the Air Ministry, the chief civil servant who effectively runs the entire department. You have had a long and much-respected career across different ministries and departments and have been the head of the Air Ministry since 1939. Despite dealing with many crises during the war, the Butt Report now threatens to be hugely damaging to your department. You head to a meeting with the political head of the Air Ministry, Secretary of State for Air, Archibald Sinclair, full of trepidation.

The Butt Report is damning indeed. Butt argued that in reality, and despite the claims of the bomber crews, only around 20 per cent of

MOST SECRET

SUMMARY STATISTICAL CONCLUSIONS

An examination of night photographs taken during night bombing in June and July points to the following conclusions:

1. Of those aircraft recorded as attacking their target, only one in three got within five miles.

2. Over the French ports, the proportion was two in three; over Germany as a whole, the proportion was one in four; over the Ruhr, it was only one in ten.

3. In the Full Moon, the proportion was two in five; in the new moon it was only one in fifteen.

4. In the absence of haze, the proportion is over one half, whereas over thick haze it is only one in fifteen.

5. An increase in the intensity of A.A. fire reduces the number of aircraft getting within 5 miles of their target in the ratio three to two.

6. All these figures relate only to aircraft recorded as attacking the target; the proportion of the total sorties which reached within five miles is less by one third. Thus, for example, of the total sorties only one in five get within five miles of the target, i.e. within the 75 square miles surrounding the target.

Source I: The Butt Report

bombers were in fact getting within even five miles of their targets, a proportion that got worse when attacking the industrial Ruhr region. Full moonlight aided accuracy (though it increased losses), but poor weather and a new moon lowered the figure still further. This was

3. Arthur Street

hardly precision bombing, and in truth meant that the heavy sacrifices of RAF Bomber Command against German industrial targets had so far contributed little to the Allied war effort. And Bomber Command had lost well over 600 aircraft by the early summer of 1941, a loss rate of 15 per cent of aircraft on each mission. This was sobering indeed. What had been the point of it all?

It was all a far cry from the call to arms back in July 1940 when Churchill had announced to Lord Beaverbrook that the bomber fleet would be the route to success (see Source 2).

When I look round to see how we can win the war I see that there is only one sure path. We have no Continental army which can defeat the German military power. The blockade is broken and Hitler has Asia and probably Africa to draw from. Should he be repulsed here or not try invasion, he will recoil eastward, and we have nothing to stop him. But there is one thing that will bring him back and bring him down, and that is an absolutely devastating, exterminating attack by very heavy bombers from this country upon the Nazi homeland. We must be able to overwhelm them by this means, without which I do not see a way through.

Source 2: Churchill to Beaverbrook, 8 July 1940

The Prime Minister and the Chiefs of Staff had backed the bombing offensive as part of a wider strategy for tilting the war back in Britain's favour. But it has not gone well, despite the enthusiasm of Churchill for the bombing campaign. Bomber Command had (and still has) many weaknesses in equipment, technology and tactics, but the bombing offensive was further undermined in 1940 by the demands of national survival. The RAF's efforts were focused on holding off the Luftwaffe in the Battle of Britain (in which Bomber Command played a crucial, if too-often-overlooked role) and then in supporting the Royal Navy in protecting Britain's maritime supply routes in the Battle of the Atlantic (a role in which the RAF, this time Coastal Command, played a pivotal and often-forgotten part).

The greatest obstacle to success beyond these distractions, however, remains the enormity of the task confronting Bomber Command. Indeed, although Churchill and the Chiefs of Staff had agreed to maintain and then ramp up the bomber offensive in mid-1940, the approach was still to focus on precision attacks against key economic or military targets, principally synthetic-oil production. Support for the bombing offensive had been bolstered by the promotion in the summer of 1940 of Charles 'Peter' Portal to the post of Chief of the Air Staff, a position from which he has been able to back the bombing campaign to the fullest extent possible. But his replacement as Commander-in-Chief of Bomber Command, Richard Peirse, is a confirmed supporter of precision industrial bombing. Portal has tried to push him towards a broader approach that would also emphasise attacks on German morale, but Peirse has resisted; he prefers to maintain his force's fighting strength and to focus his efforts on precision targets in the best operating conditions – namely, in moonlight.

His methods have been hampered, however, by continuing difficulties in bombing accuracy, the limited range of his bombers and the inclement winter weather over north-west Europe. Results have not markedly improved and Portal, as head of the RAF, has been feeling the heat from Churchill's government. We are 'on the defensive with a vengeance' he had informed Peirse back in February 1941, urging him once more to consider broadening his range of targets to include urban centres. Peirse asked for more time for his precision-bombing

4. The disappointing Avro Manchester

campaign to show results, and suggested that a new range of aircraft now starting to come into service would turn the campaign round. Existing aircraft had also been improved, but great things had been expected of new bombers, such as the Short Stirling, Avro Manchester and Handley Page Halifax. It has not really worked out yet. The Manchester appears to be a huge disappointment, whilst the Stirling's operating altitude is reportedly too low, rendering it more vulnerable to enemy action. The Halifax shows more promise, but overall results in the bomber offensive have not been heartening. Peirse is still hoping for more time to improve results. Time, however, now appears to have run out.

You know that internally the RAF realises how disappointing the results of the bomber offensive have been, but now that Lindemann has taken an interest and the Butt Report has been delivered, the cat is out of the bag. Lindemann is a maverick, willing to challenge existing views and approaches, often in a quite confrontational and aggressive manner. He is of German-American parentage and, rather curiously, still speaks with a mild German accent. His solutions to issues are often 'left-field' and range from deeply flawed to perspicacious. Egotistical, arrogant and condescending, Lindemann irritates many of those around him, but Churchill values his interventions. As one colleague noted,

'Churchill used to say that the Prof's brain was a beautiful piece of mechanism, and the Prof did not dissent from the judgment.'

After reading the Butt Report, Churchill demands that Portal and the Air Ministry review the whole campaign with the 'most urgent attention'; Peirse remains in denial and simply cannot believe it is as bad as Butt suggests.* As you confer with your boss, Archie Sinclair, the assured and charismatic leader of the Liberal Party as well as Secretary of State for Air, you must plan on damage limitation. Under the direction of Portal, you task the Directorate of Bombing Operations to issue your own Air Ministry report on bombing operations. But what line can be taken, and how – and, indeed, should – the bombing offensive be saved in its current form? Portal remains an enthusiastic advocate of wider area bombing, with the intention of targeting enemy morale; this will be politically difficult to admit openly, but could well be the only way to maintain the offensive. Portal promises 'jam tomorrow'; he claims that with a force of 4,000 heavy bombers, strongly focused on blasting Germany's major towns and cities into rubble, enemy civilian morale would collapse.

Sinclair and you worry about whether such a vision is either palatable or, indeed, desirable. Peirse still wants to persevere with precision methods, but although that will be politically more acceptable, and potentially more strategically effective, there is no sign of it working any time soon. In contrast, Portal's policy of area bombing will provide a greater chance of securing the backing of Churchill and keeping the bombing campaign on track, but it could put at risk the drive for precision bombing. Indeed, where might a push towards area bombing end up?

In addition to suggesting that bombing operations be integrated into an overall strategy for winning the war in conjunction with the navy and the army (some in the RAF still want to argue that the war could be won by bombing alone), which approach should the RAF move towards?

---

* In truth it was worse, as post-war analysis later demonstrated. About 50 per cent of bombs simply fell in the open countryside, and perhaps as few as 5 per cent of bombers on each mission ever bombed within five miles of their targets.

AIDE-MEMOIRE

* Is the precision-bombing idea at all achievable?

* The jury is out on whether area bombing will have any material effect on the German war effort.

* The Butt Report shows that precision bombing in its current form is simply failing and is costing huge resources.

* The military and strategic benefits of precision bombing will be significant, possibly decisive, if it can be made to work.

* It will be easier to achieve real results with area bombing faster than with precision bombing. If you do not get some success soon, the whole campaign could be closed down.

* Precision bombing will avoid the moral stigma of causing widespread civilian loss of life.

* How might such an area-bombing campaign against civilians go down in Washington, where Churchill is desperately trying to curry favour with Roosevelt's administration?

Churchill pondered. Was this actually achievable? Had not Butt just reported the desperate limitations of Bomber Command – problems that would not be solved easily with a switch in policy and with yet more vast resources thrown at it? What advice should you and Sinclair offer?

THE DECISION

Should you recommend to Sinclair that he back Portal's drive for more area bombing, or should you support Peirse's view that precision bombing will work (and be more effective) if you stick with it?

▶ If you want to switch emphatically to an area-bombing offensive, go to **Section 4** (p. 248).

OR

▶ If you want to focus your resources on solving the precision-bombing difficulties, go to **Section 5** (p. 256).

SECTION 3
ALTERNATIVE TARGETS

14 May 1941, Defence Committee meeting, Cabinet War Rooms, Whitehall

Once again the Prime Minister, Winston Churchill, is chairing a Defence Committee meeting at Whitehall. One of the main items on the agenda is the employment of air power in future campaigns, particularly as Britain's immediate crises seem to have been alleviated and the threat of defeat appears to have eased. Once again the whole issue of the bomber offensive is back at the forefront of Air Staff thinking. Charles Portal has tabled a new document arguing for a much greater investment to be made in RAF Bomber Command and for the emphasis to be placed squarely on an intensified bombing campaign against German cities and industries. You are **Admiral of the Fleet Dudley Pound**, the First Sea Lord and Chief of Naval Staff. At sixty-two years old, you are a veteran of many decades of service; you even commanded a battleship at the Battle of Jutland back in 1916. Your health, however, is no longer robust and

5. First Sea Lord Dudley Pound

you have been having trouble sleeping; you have even been known to doze off in important meetings. But not this time. As you look on and listen to the unfolding debate, you start to become concerned about Portal's plan and the possible diversion of aircraft to the bombing of Germany and away from battling the U-boats in the Atlantic.

It all seems to symbolise a return to the debates of a year ago, when the RAF's bomber offensive had been reined in to divert resources to other theatres and campaigns. In part the Air Staff had acquiesced on the proviso that this was to be considered a breathing space for Bomber Command to rebuild. But the redirection of Bomber Command into other duties, and the employment in other theatres of aircraft that might have been used to bomb Germany, has proved fruitful.

The summer of 1940 had been crucial in Britain's war against Hitler's Germany. With Allies falling by the wayside and military disasters everywhere, Churchill's new government was forced into a major reappraisal of strategy. A crucial first step was to ensure survival, first against the expected assault by the Luftwaffe, and second by securing the shipping lanes into and out of the British Isles. Failure in either of these two areas would prove calamitous, but as the Prime Minister had already announced, 'Hitler knows he will have to break us in this island, or lose the war.'

Fighter Command had defended the air space over Britain, but a crucial role was also played by Bomber Command. Night after night the bombers set out to attack enemy ports, invasion forces and communication and transportation networks, all in an effort to hinder Germany's ability to mount an invasion of the British Isles. Desperate times required desperate measures and, with mounting losses on these anti-invasion missions, Churchill and the War Cabinet Defence Committee insisted that the Air Staff suspend bombing operations against industrial targets in Germany, to enable a greater effort against the mounting threat of an invasion. Losses on deep bombing operations had been crippling Bomber Command anyway, and the immediate benefit of such strikes was by no means clear. The Air Staff had resisted such moves, but they were overruled.

By the autumn the RAF had fought the Luftwaffe to a standstill, and the invasion threat had receded. But although Portal, now installed as the new Chief of the Air Staff, had pressed for a renewal of the bomber offensive against German cities and towns, another and perhaps even greater threat had emerged in the deep waters of the Atlantic. When the Germans had captured the French Atlantic ports – Brest, Saint-Nazaire, Lorient – they had been afforded much safer, more direct and closer access to the North Atlantic for their U-boat fleet. Suddenly the U-boat threat had grown to deeply worrying levels, and Allied shipping losses had rocketed out of control to quite unsustainable levels – in June 1940 the monthly losses in the North Atlantic had been 50,000 tons; by November the figure had grown fivefold.

The Admiralty began to lobby vigorously for any available long-range aircraft to be deployed over the Atlantic to help fend off the U-boat threat, but these were exactly the same types of aircraft that Bomber Command wanted for their operations against Germany. RAF Coastal Command was woefully short of aircraft, and specifically those types suitable for long-range operations over the Atlantic to combat the U-boats. Bombers were more than capable of fulfilling the role, and with the reorientation of air-power resources away from bombing for the time being, new bombers were thrust into action against the U-boats, not only in providing air cover, but also in attacking the U-boats' home bases in France. By the spring of 1941 the Battle of the Atlantic appears to have been won. Quick, decisive and repeated action against the French ports, and the inability of the U-boats to inflict heavy enough losses in the Atlantic, seems to have prompted Hitler and his U-boat commander, Karl Dönitz, to suspend the U-boat campaign.

In addition, the Germans now seem to be massing their forces for an invasion of the Soviet Union, and so with the two most direct threats to British survival apparently under control, and Hitler looking elsewhere, Portal has reignited the bomber-offensive plan. The Air Staff are pushing for limited missions against military / industrial targets, under the direction of Air Marshal Richard Peirse, a firm advocate of precision bombing. But the RAF's heavy-bomber fleet is still small, and scepticism about its ability to deliver on its promises remains high. The army wants more bombers to turn the tide of the Mediterranean campaign firmly in Britain's favour, and you lobby hard for more aircraft to secure the safety of the Atlantic shipping lanes and to support naval operations in the Mediterranean. Churchill and the Defence Committee throw a few crumbs to Portal and Bomber Command, in terms of developing technology and long-term investment, but there is no stomach for switching the available air-power resources towards the bomber offensive just yet.

AFTERMATH

When the USSR is drawn into the war in June 1941 a more conventional path to victory – rather than the theoretical grand vision offered by the Air Staff – opens up. A combined Commonwealth and Soviet war effort could well redress the balance of power in the war, if the USSR can withstand the German blitzkrieg. Success for the British in the Mediterranean would also potentially divert resources away from the Eastern Front.

Bolstered by a strengthened RAF, the British won comprehensive control of the air space over the Mediterranean and the Middle East in 1941, driving the Italian and German air forces in the theatre from the skies. With the increase in air support and the damaging effects of British air power on Axis supplies into North Africa, the tide of the North African campaign turned decisively in favour of the Allies across 1941, although the RAF could do little to stem the German and Italian offensives in the Balkans. A robust air defence of Crete nonetheless saw off an attempt by the Germans to use airborne forces to seize the island in May, with the Axis units enduring crippling losses.

Yet when the Americans entered the war in late 1941, equipped with their grand theories of strategic air bombardment of the German economy, they found little immediate support in Britain. The Air Staff and Charles Portal were enthusiastic about developing the British bombing offensive to work alongside the Americans, but Britain's investment in the required long-range heavy bombers had fallen away since 1940, and although some new aircraft such as the Avro Lancaster and the Handley Page Halifax were due to come properly into service in 1942, they were fewer in number than Bomber Command had hoped for; some were, in any case, to be diverted to the Atlantic to provide near-continuous air cover for the convoys travelling between Britain and North America.

The USA's bombing campaign soon began to struggle against the Germans' air defences, and the RAF's contribution was too small to make a significant difference; Bomber Command found out that it had a great deal to learn and put right. By 1943 the campaign was floundering. In contrast, the extra commitment by the British to other aspects of air power and to the development of larger ground forces, supported by firm control of the Atlantic and victory in the Mediterranean, made a cross-Channel invasion in 1943 viable, even to the doubters in London. Although the German air force was still a threat as it was far from beaten, such was the Allies' superiority in the air that the invasion succeeded. However, the land campaign in northwest Europe, though ultimately successful, dragged on through 1943 and 1944 and saw heavy loss of life, something the British had always wanted to avoid.

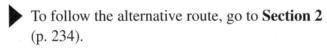

THE DECISION

▶ To follow the alternative route, go to **Section 2** (p. 234).

OR

▶ To know more about the bomber offensive, go to the **Historical Note** on p. 269.

SECTION 4
FIRESTORMS

18 January 1945, office of Charles Portal, Air Ministry, Berkeley Square, London

Marshal of the Royal Air Force, Charles Portal, Chief of the Air Staff, is facing a momentous decision. Although the end of the war appears to be in sight, following the Allies' successful D-Day invasion in Normandy in June 1944, Portal still has a major problem in that one of his senior commanders, Air Chief Marshal Arthur Harris, Commander-in-Chief of RAF Bomber Command, appears to be contradicting the Air Staff directives and Portal's wishes. Harris simply cannot agree with Portal's assessment of the bombing campaign so far, and therefore with where the priorities must be placed in current bombing policy. And now Portal has just received a letter in which Harris threatens to resign, if Portal wants him to.

You are *Air Vice Marshal Norman Bottomley*, the Deputy Chief of Air Staff (DCAS), and Portal has called you in to discuss the Harris situation. You have been involved in military aviation since the First World War and have held a series of increasingly senior positions, culminating in your appointment as DCAS in 1941. You had been a firm advocate of area bombing in 1941, but since the successful liberation of France and the growing effectiveness of precision bombing techniques in Bomber Command, you have adapted your thinking. Like your boss, Portal, you now had doubts about the necessity and effectiveness of area bombing; much had changed since the dark days of the bomber offensive back in 1941.

Although Churchill had backed the increasing use of area-bombing methods as a way of keeping the bomber campaign going in 1941, little appeared to have improved by early 1942. Following the impact of the Butt Report, Air Marshal Richard Peirse, Commander-in-Chief of Bomber Command, had shifted his tactics towards easier but lower-tariff targets, but German opposition had strengthened yet further. The enemy's Kammhuber radar-defence guidance network superbly aided the efforts of the Luftwaffe's specially equipped night-fighters to create a formidable obstacle to Bomber

6. Air Vice Marshal Norman Bottomley

Command. Increasing losses, sceptical questioning and declining support in the newspapers pushed the campaign to the point of collapse.

Portal's patience then ran out. In February 1942 the Air Staff issued a new directive to Bomber Command, which put the targeting of Germany's towns and cities squarely at the centre of RAF policy. The central aim was then to break the 'morale of the enemy civil population and in particular of the industrial workers'. High-value economic targets – synthetic-oil production was still on the list – would continue to be attacked, but the straightforward approach of flattening Germany's urban centres, in order to break the will of the workers and thus cripple the German economy, was increasingly the cornerstone of the RAF's thinking.

Who was to drive this new strategy forward? Peirse remained at heart an advocate of precision bombing and thus was not the man for the job. He was moved on to other pastures and eventually achieved social notoriety: while in India he began a scandalous affair with General Claude Auchinleck's wife, causing Peirse and Lady Auchinleck to be sent home in disgrace.

Portal had decided that a firm believer in area bombing was required for Bomber Command and chose Air Marshal Arthur Harris to take over. Known in the popular press as 'Bomber Harris', but to

7. Air Marshal Arthur 'Butch' Harris

his crews as Butch – short for 'Butcher', after the heavy casualties that he was willing to accept – Harris was not the architect of the area-bombing policy, as this was an Air Staff directive issued to Bomber Command before his appointment. Yet he was an enthusiastic advocate of the policy. He had few qualms about bombing Germany. 'They sowed the wind and now they are going to reap the whirlwind,' he declared. Tough, obdurate and determined, Harris brought a firm grip and single-minded purpose to a command that in early 1942 desperately required such leadership.

As a commander-in-chief, Harris refused to be blown off-course, as perhaps Peirse had been. Harris had a clear view of how his command should operate and he stuck to his guns. Although he received much judicious advice about crucial economic targets that, if destroyed, would quickly cripple the German economy, he remained sceptical about what he described as 'panacea targets' – he stuck fervently to the policy of area bombing as the most effective route to victory.

Bomber Command began to achieve better results, partly due to Harris' drive, and partly because of the introduction of new and improved technology. By the summer of 1942 the famous Avro Lancaster bomber had begun to arrive in service, an aircraft that would become the linchpin of Bomber Command for the rest of the war. It was Harris' best heavy bomber, capable of carrying a substantial payload of bombs deep into Germany. Bomber Command also began to take delivery of new Gee electronic-guidance equipment, which improved the effectiveness of operations, but was only suitable for short- to medium-range raids. In 1942 Bomber Command had scored some notable successes, including the high-profile 1,000-bomber raid

8. Avro Lancaster

on Cologne, but as the year went on, results tailed off in the face of determined German opposition and various technical difficulties.

By late 1942 the US Army Air Forces (USAAF) had deployed to Europe, and they also subscribed to the idea of using heavy bombers to attack economic and industrial targets. Despite Harris' warnings, the Americans had been determined to prosecute a daylight precision campaign, similar in many ways to the RAF's approach in 1940, but backed by much greater resources. They believed that precision bombing was possible with sustainable losses, and would hit the Third Reich harder and more precisely than the bludgeoning blows of Harris' area-bombing campaign.

By 1943 the Allies were 'bombing round the clock', with the Americans doing so in daylight and the British at night, but although it appeared to be a coordinated combined offensive, it was not; the two air forces did their own thing, believing the other to be misguided. The Americans went through a very similar and painful learning curve to the British, and it was not until the spring of 1944 that their approach began to pay off, and only because they had the capacity to take heavy casualties; the USAAF suffered appalling losses in the autumn of 1943. But in the winter of 1943–4 they had introduced long-range fighters to sweep ahead of the bombers, in an effort to cripple the

Luftwaffe and open the skies up for the bombers – in effect, to win command of the air. The Americans had been willing to take on the Luftwaffe in a brutal, short, casualty-intensive campaign to seize air supremacy in a way the RAF could, and would, not.

Harris' and Bomber Command's policy had remained one of evading the Luftwaffe and sticking to night-time area raids. Debate grew, however, as to whether the British should switch back to precision bombing, as this might yield better results. Harris generally blocked such ideas, stating that he had heard all this futile economic theorising before and that he was unwilling to commit his 'old lags' to operations that he did not believe would make that much difference. And Harris delivered spectacular and awful success in the summer of 1943 over Hamburg. In a series of heavy raids in which a large number of incendiary bombs were dropped, and because of the dry, hot weather, Allied bombing created a firestorm in which temperatures reached 1,000 degrees centigrade and caused some 40,000 people to perish.*

But there were also setbacks. A protracted bombing campaign against Berlin in 1943–4 had largely failed, and Bomber Command had suffered heavily in trying to make it work. Harris had told his crews, 'Tonight you go to the big city [Berlin]. You have the opportunity to light a fire in the belly of the enemy which will burn his black heart out.' Instead the RAF had suffered grievously, though Berlin too had endured appalling losses.

By the summer of 1944 much had changed. The Luftwaffe had suffered such crushing losses against the Americans in the spring of that year that they had effectively been finished as a major threat to the Allies. The Allied bomber forces were then drawn into supporting the D-Day landings and the campaign in France for much of the summer of 1944, and it was not until September that Harris had been freed to continue his bombing campaign against Germany.

But what should Bomber Command target? And what methods should they use? The RAF now had at its disposal technology that

---

* Albert Speer, Hitler's Armaments Minister, claimed that six such raids in quick succession might have ended the war, although this was of course difficult to prove. Bomber Command was in any case incapable of replicating the attack, to order.

supported precision-bombing techniques, equipment that had been designed to aid night-time bombing but which, other than in the clearest of weather – something of a rarity in Northern and Central Europe – actually now proved more accurate than the Americans' daylight bombing tactics. Harris had had two ideas thrust upon him. Portal had reverted to the RAF's 1940 idea of destroying Germany's synthetic-oil plants, a strategy made all the more viable by the denial of the Romanian oil fields to Germany, as they had been captured by the Soviets. In contrast, Air Chief Marshal Arthur Tedder, Eisenhower's deputy commander for the Allied forces in Europe, preferred to attack the German transportation network. This, he argued, would paralyse the German state and facilitate a rapid Allied victory. Both cases had merit.

Yet Harris demurred. He still believed that area bombing was the shortest route to victory and that changing course now merely invited confusion and dispersal of effort, just as victory was in sight. He remained unconvinced that the precision methods Bomber Command had been developing would prove that successful in a 'panacea'-style campaign against oil – Bomber Command had been down this road before.

There was also a growing voice of opposition to the bombing of civilians. There had always been dissent on religious, moral and even strategic grounds – Bishop George Bell had previously criticised the government for undermining the moral case for Britain's war effort by indulging in barbaric bombing, and there was an ongoing 'Bombing Restriction Campaign'. By 1944, especially after the success of D-Day and the collapse of German control of the Balkans and much of Eastern Europe, some people were more openly and vigorously expressing their distaste. Vera Brittain's pamphlet *Seeds of Chaos: What Mass Bombing Really Means* took on Bomber Command's claims and attacked the popular press for blithely going along with the bombing of civilians.

Portal, you and the rest of the Air Staff eventually settled on a policy that prioritised hitting oil production, followed by transportation targets. If there was nothing else, due to weather or operating conditions, Bomber Command was released to attack area targets. Harris appeared to have grudgingly accepted Portal's directive, but he continued to grumble. There had also been rumours that he was evading the intent of the Air Staff directive and was reverting to his own area-bombing techniques all too readily. Even though there were, in your view, clearly positive effects of the oil-targeted bombing campaign, Harris continued to be obdurate and obstructive, questioning the choice of targets. Portal had begun to wonder whether Harris was really obeying his orders.

Throughout the autumn of 1944 Harris and Portal had been arguing over the best bombing strategy for Bomber Command, and although Portal was Harris' boss, Harris would not give way. Now Harris has written to Portal again, contesting the Air Staff's policy. Yet neither you nor Portal can be sure that Harris has flagrantly disobeyed Air Staff orders; he has continually argued that Bomber Command has only reverted to area bombing when other priority targets have not been available, due to the weather. You suspect that Harris has been dropping the oil and transportation targets far too readily, but can you be sure? Ultimately, if Harris is not willing to prosecute the Air Staff's bombing policy as fully as possible, should Portal call his bluff and accept his resignation?

AIDE-MEMOIRE

* Portal and the Air Staff decide policy; Harris puts it into action. Does he accept this?

* It appears to some at the Air Ministry that Harris is displaying rank disobedience.

* Harris has carried out much of what had been asked of him, however grudgingly. Can you be sure he is actually disobeying Air Staff directives?

* Any other commander who appeared to be disobeying or evading orders might well have been sacked by now.

* What would sacking Harris – one of the most well-known, high-profile Allied commanders – do for Bomber Command morale?

* Harris had been an admirable firm, determined leader in 1942, but now his unwavering fixation on area bombing appears to be undermining Bomber Command's credibility.

* Is now the time to take the foot off the pedal? Germany's urban centres are being systematically reduced to rubble, and this is surely keeping the pressure on Hitler's regime.

THE DECISION

What should Portal do? He asks you for your advice.

▶ If you want to sack Harris, go to **Section 6** (p. 262).
   OR
▶ If you want to let him carry on, go to **Section 7** (p. 266).

SECTION 5
PRECISION AND STEALTH

14 November 1942, RAF Bomber Command HQ, High Wycombe

At Bomber Command HQ a group of elite Allied air force officers are assembling to discuss the bombing offensive against Germany. Operational commanders and their teams are all steadily arriving and are assembling alongside United States Army Air Force personnel from the US 8th Bomber Command, which has also begun small-scale operations against Germany. Heading the conference is Air Chief Marshal Richard Peirse, Commander-in-Chief of RAF Bomber Command, with the American Brigadier-General Ira Eaker, commander of US 8th Bomber Command, acting as deputy. Air Chief Marshal Charles Portal, Chief of the Air Staff, is also in attendance. You are *Air Commodore Sidney Bufton*, the thirty-four-year-old Director of Bomber Operations and a driving force in pioneering new tactics and methods intended to enhance bombing accuracy and policy. After holding command positions in the early years of the war, you were moved to Bomber Command HQ to lead the development of precision bombing in 1941. Your most significant contribution has been the introduction of a specialised unit of aircraft that can more accurately mark the target for the following bombers to attack. Your Pathfinder Force received the enthusiastic backing of Peirse and Portal, though Arthur Harris, the Deputy Chief of the Air Staff, was openly sceptical. The Pathfinders have, however, proved very successful since their introduction in the spring of 1942 and your reputation has been greatly enhanced.*

Now, however, the future of the Allied bombing campaign is going to be thrashed out at Peirse's Bomber Command conference. What are its prospects? The Butt Report had almost put Bomber Command out of business in 1941, and Portal's plan for a 4,000-strong heavy-bomber fleet remorselessly demolishing German cities and towns in an area campaign had not won much support from Churchill's government, either. Portal had been

9. Air Commodore Sidney Bufton

instructed to continue with the precision-bombing campaign against easier targets and to work towards solving the problems Butt had identified. Portal had been dismayed, but was forced to accept the strategy. He eased off on the pressure that he had been applying to Peirse about area targets; the focus was to remain on lower-risk precision attacks. This was much easier to demand than to achieve, however. The underlying difficulties remained and would not be easily solved until new aircraft, techniques and technology were introduced.

Key events in the winter of 1941–2 had begun to alter the seemingly intractable problems confronting Bomber Command. First, Peirse had been instructed earlier in 1941 to attack the U-boat bases being constructed at a number of French ports on the Atlantic coast, and he now saw them as objectives that were perhaps more easily targeted and destroyed. Consequently, Bomber Command had conducted an intensive and sustained campaign to hinder or prevent their completion. Coastal targets provided less of a navigational challenge and were initially less well defended than German industrial targets.

---

* Historically, the Pathfinder Force was blocked by Harris until August 1942.

Although there were qualms about hitting French civilians (they were not the target, but many would be caught up in the bombing raids), the campaign was supported by the Air Staff and by Churchill, who recognised that success stories were required. The campaign went well, and considerable damage was caused both to the port installations and to the U-boats themselves. Bomber Command losses were not inconsiderable, but lessons were learned and a degree of success achieved.

The second major development had come on 7 December 1941, with the entry of the USA into the war following the Japanese attack on Pearl Harbor. The US Army Air Forces had their own carefully thought-out strategy for a bombing campaign – specifically, a precision daylight offensive, quite similar to the British concept of 1940 – one that the RAF now thought quite unworkable. The British were convinced that daylight raids were too risky and that night-time raids using technology-assisted tactics were the way ahead. The Americans, however, had stuck to their thinking that daylight raids with heavily armed, self-defending bomber formations offered the best approach. Their 8th Bomber Command forces have been building up for some months, and a few weeks ago they began tentative operations.

The third development centred on emerging new technology and tactics that the Air Staff hoped would begin to enhance accuracy and navigation. A new aerial blind-bombing system based on the use of radio signals, codenamed Oboe, had been introduced in early 1942 and it dramatically enhanced navigation and bombing accuracy. In early operations against targets in northern France accuracy was improved in night-time raids, such that 50 per cent of bombs fell on the target, as opposed to the much lower figures of 10 per cent identified in the Butt Report. In further raids on the U-boat bases in 1942 Oboe proved to be a major asset. It was, however, still only a partial solution, as the system was reliant on ground-based transmitters and equipment to guide the bombers; as the attacking aircraft flew further away from the guiding bases and disappeared over the horizon, the system stopped being effective. Targets in Germany were simply too far from Britain for Oboe to be a total solution to the issue of precision bombing.

Yet further possibilities existed. Although most bombers did not fly at high enough altitudes to make use of Oboe deeper into Germany, the newly introduced De Havilland Mosquito aeroplane could, meaning that greater bombing accuracy was achievable right into the heartland of the Ruhr industrial area. The Mosquito's frame was constructed mostly of wood and the aircraft was powered by two Merlin engines, making it extremely fast and able to match the speed of the German interceptors it was up against. It could also carry a payload of bombs heavier than that of the American B-17 Flying Fortress bomber now arriving in Britain. Development and production of the Mosquito have been hastened since mid-1941, following the frustrations of the early bomber campaign, and now, by the autumn of 1942, more of them are arriving in service. They provide a radically different bombing option to the more usual four-engined heavy-bomber concept.

Under Peirse's leadership, Bomber Command has started employing sneak daylight bombing tactics using the Mosquito force, and the results seem positive. Flying low to evade detection until it is too late, and using speed and manoeuvrability to survive, such attacks against very specific smaller targets have proved effective, although at a cost. The training and skills required to make the missions work have required a major investment by Bomber Command; and sceptics remain. Overall results are impressive but limited in scope; there are simply not enough Mosquitos, trained crews and pieces of high-tech

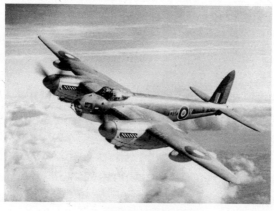

10: De Havilland Mosquito

radio equipment to drive home the possibilities of these tactics and methods.

The conference at Bomber Command proves successful. You manage to press the idea of expanding your Pathfinder Force, and Eaker and Peirse's staff and leaders agree on a number of initiatives and approaches. The concept of a combined offensive with the Allied nations working closely together takes shape. The American and British-led bombing campaign will employ American daytime raids with heavy bombers, British daylight raids with the fast Mosquito-force concept, and a night-time bombing campaign spearheaded by the RAF. Arthur Harris' efforts to push an area-bombing campaign onto the agenda come to nothing.

AFTERMATH

By the closing stages of 1942 the Germans began introducing countermeasures to the stealth fast bomber raids, using highly trained and expert fighter-pilot units to intercept the Mosquito forces; British losses began to mount. The use of the Pathfinder Force provided an enhancement to night-time bombing missions being carried out by the heavy-bomber fleet, but success was still limited, due to the reduced investment in the heavy bombers. The Americans soon encountered the realities of prosecuting daylight raids over Germany – initial losses were heavy and tactical difficulties remained, issues that had driven the RAF to night-time bombing raids two years earlier. The Americans remain determined, however, and Ira Eaker and his staff are convinced that the only way to win the air war is to attack the Luftwaffe at source.

At the Casablanca Conference in early 1943 the Allies commit their bombing forces to a combined offensive, bombing around the clock. Peirse and Eaker work closely together to link their various efforts, but German resistance remains a problem, and the results do not markedly improve until the autumn of 1943. Allied losses are still worryingly high, and there is no doubt that bombing alone is not going to bring Germany to its knees.

THE DECISION

▶ To follow the alternative route, go to **Section 4** (p. 248).

OR

▶ To know more about the bombing offensive, go to the **Historical Note** on p. 269.

SECTION 6
THROTTLING BACK

24 January 1945, Defence Committee meeting, Cabinet War Rooms, Whitehall

Churchill's Defence Committee is about to consider once again British bombing policy against Germany. Much has happened in the last week, and as Chief of the Air Staff, Charles Portal, begins to discuss the situation, he has you by his side to allay fears and concerns about the leadership of Bomber Command. You are *Air Vice Marshal Norman Bottomley*, the newly appointed Air Officer, Commander-in-Chief, Bomber Command.* Your elevation to the position has come as something of a shock. When on 18 January 1945 Air Marshal Arthur Harris had offered to resign if Portal did not believe him to be doing his job properly, it is doubtful he really expected to be taken up on the offer. But you witnessed that Portal's patience had run out completely. He and Harris had been hammering away at each other for weeks, with Harris openly contradicting Portal's and the Air Staff's agreed bombing policy. He had to go. Portal had asked if you would step in to take charge of Bomber Command and fulfil the intent of the Air Staff directives. You agreed. Once you had been an advocate of area bombing but, like Portal, you are now on board with a more nuanced approach to bombing: area bombing when absolutely necessary, but otherwise a campaign against oil and transportation is now much the better option. Portal conjured up the courage to accept Harris' offer to

---

* A post that, in reality, he took up in September 1945.

resign and, despite the shock and bewilderment that his removal generated, Churchill and his government accepted the decision.

As the Defence Committee begins to discuss the bombing campaign, matters are complicated by the latest report, *German Strategy and Capacity to Resist*, prepared for Churchill by the Joint Intelligence Committee. The authors claim that Germany is moving large numbers of units to the Eastern Front to hold back attacking Soviet armies; if this transfer could be stemmed, the Soviets might force a collapse of Hitler's defences that would bring the war to an end in April; otherwise, the war might drag on into the autumn. What, Churchill asks, of the much-discussed Operation Thunderclap, the plan to heavily bomb Berlin and Germany's eastern cities as a shock-measure?

Bombing has moved on, Portal counters. You propose that the best way to intervene is to continue attacking key points in the enemy's transportation network and ending German oil production. Both measures would serve to stymie any movement of enemy forces around Germany. Churchill demands an immediate plan of action from you.

Over the next day your team identifies crucial chokepoints in what remains of Germany's crumbling transportation network, alongside oil plants that will also continue to be a priority. However, little consideration is given to area bombing; it will only be used if the weather forces you away from other useful targets. Portal later adds that a major blitz-style air raid against a German city in the east might aid the Soviet advance, but such a mission should in no way compromise the bombing of oil and transportation targets. Churchill grumbles that he wants a show of strength – 'basting the Germans as they retreated from Breslau,' as he puts it – but for the moment he is dissuaded by Portal and the Chiefs of Staff.

AFTERMATH

The Bottomley Plan is adopted and prosecuted with vigour and produces excellent results. Bombing against the priority targets proves fruitful and, with enemy opposition to bombing almost non-existent, all goes well. By mid-February many of the main targets have been demolished, and although many repeat missions have to be carried out to keep transportation points continually suppressed, the effects are noted as being highly effective.

Bombing of towns and cities continues at times, and Chemnitz in eastern Germany is hit hard in mid-February. At the end of the month the Americans initiate Operation Clarion, in which their air forces spread out across what is left of the Third Reich and impose a near-continuous aerial threat to paralyse all enemy movement. The RAF adds its own weight to the plan, and Bomber Command switches to daylight raids as well as night-time bombing to increase the pressure. Large-scale area bombing of urban centres is finally suspended in early March. The Air Staff have always been concerned about how their long campaign would be viewed in the post-war world and they do not want to act provocatively as the war draws to a close. Although dissenting voices continue until the final moments of the war in Europe, the RAF negotiates the final months with its reputation intact, and with a grudging acceptance that Bomber Command has made a valuable and costly contribution to Allied victory.

11. Bombed Chemnitz, 1945

THE DECISION

▶ To follow the alternative route, go to **Section 7** (p. 266).

OR

▶ To know more about the bomber offensive, go to the **Historical Note** on p. 269.

SECTION 7
DRESDEN

13 February 1945, Dresden, Germany

On the night of 13 February 1945 800 RAF heavy bombers begin pummelling Dresden, a city that has little direct military importance, but one through which many refugees, civilians and troops are transiting. Most people are fleeing westwards to evade the clutches of Stalin's Red Army. The city had been identified as a possible target for a major raid a few weeks earlier, as part of an ongoing intensive campaign to compel Germany's surrender. There are also hopes that such an attack will aid the Soviet advance through eastern Germany. After the RAF finishes its bombing, an intensive raid by the Americans follows the next day, loosely targeting the city's railway marshalling yards, although in truth this is a broad area attack. Because of the weather conditions, the intensity of the bombing, the heavy use of incendiaries and the combustible nature of many buildings in Dresden, the attack causes a firestorm to be generated, in which some 30,000 people lose their lives, although the figures vary, with some estimates being very much higher. Most of those killed are clearly non-combatants.

Over the next few days it starts to become obvious that the Dresden raid, with the firestorm and the appalling loss of life, is shifting the balance of the debate on the general policy of bombing German towns and cities. World opinion begins to turn against the Allied air forces, and in particular against Bomber Command. The Americans quickly distance themselves from the bombing of urban areas, arguing that they have always targeted specific military facilities. British and American news outlets begin to question the intensity of the raid – 'unprecedented fury', 'a deliberate terror bombing' and 'ruthless' are all terms being bandied around.

You are **Air Vice Marshal Robert Saundby**, Arthur Harris' deputy Commander-in-Chief of Bomber Command, a post you have held since early 1943. You had joined the Royal Flying Corps back in the First World War and had remained in the RAF throughout the interwar years. In the Second World War you had worked for Richard Peirse and, since February 1942, for Arthur Harris. You had always been a keen supporter of the bombing campaign and have backed Harris throughout the last three years. Now, however, you begin to worry about the impact of the negative publicity surrounding the Dresden raid. You try to advise Harris about his language and hard-line stance, but to little effect.

Harris is also starting to become dismayed. Back in January he had faced down the Chief of the Air Staff, Charles Portal, throwing down the gauntlet in a 'back me or sack me' grand gesture. Portal had blinked and had then given up trying to persuade his errant commander to change his tone and tack. In a reply laced with resignation and despair, Portal had written: 'I am very sorry that you do not

12. Robert Saundby

believe in it [the plan to bomb oil and transportation targets ahead of area targets] but it is no use craving for what is evidently unattainable. We must wait until the end of the war before we can know for certain who was right.' It was a major climbdown, and Harris saw it as a vindication of his uncompromising approach.

Harris considered himself free to maintain his area-bombing campaign when he thought it prudent and the conditions allowed. The Soviet Union's advance on the Eastern Front had been the major Allied effort in the first few months of 1945, and the Western Allies had considered how best they could support Stalin's Red Army. The British Joint Intelligence Committee identified that the Germans had been shipping troops and equipment eastwards to strengthen their defences against Stalin's forces, prompting Churchill to ask how the RAF could help prevent, or at least hinder, this. An initially cautious response, which emphasised that hitting oil and transport objectives would help the most, was brushed aside by the Prime Minister; he wanted more. Harris stepped in and offered to target more major towns and cities in eastern Germany to up the pressure on the enemy: Dresden was one of those targets so listed.

Although Dresden was only one of a number of urban centres targeted by Bomber Command, the intensity of the raid and its consequences were startling. You wondered whether the press attention might create a problem. The upper echelons of the Allied leadership, however, are also beginning to get nervous, and they recognise that such destruction is perhaps undermining the Allies' moral high ground – a position that had previously seemed assured after the atrocities of the Germans' concentration camps (let alone the full horror of the death-camps) had been fully realised in the West. Now attacks on fleeing civilians in an appalling conflagration might be seen as dragging the Allies into a similar mire as the Nazis.

AFTERMATH

Churchill eventually broke cover and wrote to Portal asking if bombing German cities 'merely for the sake of increasing terror' was worthwhile.

He wondered if the destruction of Dresden constituted 'a serious query against the conduct of Allied bombing'. On one level he was correct, for the raid appears to have tarnished the reputations of Bomber Command and Arthur Harris. Yet Portal and Harris are apoplectic about Churchill's politically motivated attempt to distance himself from the awful implications of area bombing. They are convinced that Churchill is guilty of breathtaking mendacity; after all, it was he who had pressed for a more determined and aggressive use of bombing to aid the Soviet advance and to make a statement.

Portal demanded that Churchill temper his language, which the Prime Minister accepted. Nonetheless, Harris did not help himself by continuing to argue in support of area-bombing attacks. His much-quoted statement that he did not 'regard the whole of the remaining cities of Germany as worth the bones of one British grenadier' was hardly likely to soften attitudes towards him, not after Dresden.

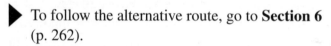

THE DECISION

▶ To follow the alternative route, go to **Section 6** (p. 262).

OR

▶ To know more about the bomber offensive, see the **Historical Note** below.

HISTORICAL NOTE

The bombing of Germany during the Second World War remains one of the most contentious issues of the entire conflict. As many as 500,000 German civilians perished as a result of Allied bombing, mostly at the hands of RAF Bomber Command. Although some people question the strategic value of the campaign, there is little doubt that the bombing offensive seriously damaged Germany's ability to wage war, though quite often not in the ways that Allied planners had originally intended. What is not in doubt is the vast investment made by the Allies in their strategic bombing fleets, an investment that was much greater than the enemy resources allocated to resisting it.

Could those Allies resources have been used more effectively? The historical pathway through this chapter was to 'take the gloves off' in the summer of 1940 (from Section 1, go to Section 2) and then to embrace area bombing as the most likely method to sustain RAF Bomber Command during the dark days of 1941–2 (from Section 2, go to Section 4). In January 1945, despite Portal's reservations about Harris' conduct of the bombing campaign, Portal baulked at sacking him (from Section 4, go to Section 7) and Harris remained in post and was thus left accountable (unfairly so) for the infamous Dresden raid a month later.

Were the other options and choices that could have been made in this chapter either viable or realistic? There is much conjecture over what might have happened if the British had abandoned heavy bombing as a central pillar of their grand strategy in 1940 (from Section 1, go to Section 3). There would probably have been greater and more rapid success in campaigns such as the Atlantic and the Mediterranean, and resources could have been diverted to ground and naval support aircraft. It would also have allowed for the creation of a general-purpose bomber force, somewhat akin to the Luftwaffe in 1940–41, though with much deeper roots and capacity.

But there would have been a price to pay. The historical bombing campaign forced the Third Reich into diverting vast resources to the defence of Germany itself. With a reduced threat, more troops and equipment would have been deployed elsewhere. If the British only had a limited commitment to the bombing campaign, could – and would – the Americans have persisted? They were almost forced to concede in the autumn of 1943, and that was when Bomber Command was fully engaged. Without the RAF drawing some of the fire, the Americans might well have suspended their efforts.

The most significant consequence of the Allies not pressing a bomber offensive would have been the continuing presence of the Luftwaffe as a viable fighting force. The bomber attacks forced the Luftwaffe into defending German air space and being drawn into an attritional battle they could not survive. It was this that eventually broke the Luftwaffe in the spring of 1944 and paved the way for the Allies' cross-Channel invasion in June 1944. Although the defeat of the Luftwaffe was principally a result of American air-power operations in 1943 and 1944, it was still in part based on British commitment in pinning down German air forces in the night-bombing campaign; without Bomber Command's sacrifices, the overall campaign might well have failed.

A great deal of speculation has taken place since the Second World War about the viability of employing precision bombers such as the Mosquito, on a much grander scale than occurred historically (from Section 2, go to Section 5). The proponents point to how such aircraft, either in an enhanced 'Pathfinder' role or even more on stealth bombing missions, could and would have reshaped the bomber offensive. Such a strategy, it is contended, would have been much more effective in achieving the key goals of attacking economic and industrial targets, as well as crucial military objectives, but would also have avoided mass civilian casualties.

Yet the vision passes over some very important issues. Most obviously, the lack of suitable aircraft, sufficient trained crews and appropriate technical equipment were massive obstacles to overcome. In addition, Britain's aircraft industry had invested massively since the late 1930s in the development of a heavy bomber fleet, and to switch quickly in 1941 to high-speed bombers such as the Mosquito, and to mass-produce them quickly enough to salvage the bomber offensive, seems unlikely. The radio technology to enhance the fleet was also only just emerging by 1941–2 and was never likely to be available in large enough numbers to increase the campaign sufficiently quickly.

Perhaps more importantly still, the notion of the fleet of fast precision bombers decisively turning the bomber campaign without a significant reaction from the German defenders seems fanciful. If the precision-bombing campaign had started to bite, the Luftwaffe would have shifted its tactics and deployments to meet the challenge.

And what of the idea of sacking Harris in January 1945 (from Section 4, go to Section 6)? The likelihood of Portal making such a move in January 1945 was quite low, but might have been appropriate; Harris was doing all he could to wriggle out of the orders the Air Staff had issued him a few months before. Yet Bomber Command still significantly ramped up its bombing of oil targets, even though post-war analysis demonstrated that the effort could have been ramped up by another 18 per cent. Whether Harris disobeyed the letter of the orders is, therefore, debatable – quite possibly he did not – but he did fulminate against the intent, and may have pushed the boundaries of his orders.

It is possible, however, that if Harris had been sacked in January, the calamitous bombing of Dresden on 13–14 February might have been avoided. Without Harris in command, Portal and Bottomley might have interpreted the intent of Churchill's desire for a big show of strength somewhat differently, and might well have looked at alternatives to city bombing with more determination.

Churchill always praised **Lord Beaverbrook**'s contribution to the war effort, particularly during the pivotal period of 1940–41. Some have since questioned how important he actually was to the expansion of the RAF, but he certainly raised the profile of the aircraft industry. It was Beaverbrook, for example, who issued the appeal for pots and pans to be handed in by the population to help build Spitfires, although this may well have been a propaganda exercise.

Arthur Street was a well-respected senior civil servant for many years and played an important role in a number of key departments after the war, following his stint at the Air Ministry from 1938 to 1945. He was also a rare bird in the upper echelons of the Civil Service: someone who had risen through the ranks from humble beginnings.

Dudley Pound's ill health hindered his ability to influence the Chiefs of Staff from time to time, and by mid-1943 he was failing fast. He passed away from a brain tumour in October 1943 a month after resigning.

Norman Bottomley, though he had earlier been a firm supporter of area bombing, repeatedly clashed with Arthur Harris in 1944 over Harris' obdurate refusal to adopt new precision methods. Nonetheless, Bottomley replaced Harris as C-in-C of Bomber Command in September 1945. He later served in a senior role at the BBC.

Sidney Bufton fought a long and tortuous campaign to introduce the Pathfinder Force, against the wishes of Harris and many of his operational commanders. Bufton was proved correct and he later served in a series of key roles in the post-war world, before eventually retiring in 1961.

Robert Saundby was a firm and staunch ally of Harris between 1942 and 1945, serving him loyally. He was forced to retire from the RAF in 1946 due to poor health, but later became a noted lepidopterist and author.

6

CASABLANCA:
ROOSEVELT AND CHURCHILL'S STRATEGIC CHOICES, 1943–44

SECTION 1
THE CONFERENCE

16 January 1943, fifty-eighth meeting
of the Combined Chiefs of Staff,
Anfa Hotel, Casablanca

It is the 'Symbol' Conference, a meeting of the Allied leaders to determine, among other things, the future direction of the war. The conference is taking place in Casablanca, Morocco, until recently part of Vichy French North Africa, but now liberated territory. It is the morning of the third day of the conference and the military Combined Chiefs of Staff are engaged in a crucial meeting. General Alan Brooke, the British Chief of the Imperial General Staff, has just begun to outline the United Kingdom's thinking for the future direction of the war against the Axis powers. The tension rises as he expertly and resolutely discusses the British position – a position that is clearly at odds with that of their American counterparts, led by General George Marshall. Around the conference table sit the senior commanders of the Allied armed forces fighting on land, at sea and in the air, from the Mediterranean to the North Atlantic and across the vast Pacific Ocean. These ongoing discussions at the Anfa Hotel in Casablanca are effectively deciding the path to victory that the Allies will take over the next two years. It is a pivotal moment in determining the outcome of the Second World War.

You are *Field Marshal John (Jack) Dill*, Chief of the British Joint Staff Mission based in Washington, and London's main representative on the Allies' Combined Chiefs of Staff, which is based in the USA and runs the military war effort. Prior to taking up this appointment at the end of 1941 you had been Chief of the Imperial General Staff, but you had clashed with Churchill too many times, particularly when, as you saw it, you had to block his more outlandish notions and ideas. Yet in Washington you had played a vital role as the link between the American and British leaderships; although British, you are highly regarded by the Americans and get on very well with Marshall in particular. Now, at Casablanca, your diplomatic skills are going to be sorely tested, for it is clear that Brooke and Marshall have sharply divided views on what path the Allies should now take.

By the autumn of 1942 the Allies were beginning to see emerging a path to victory against the Axis powers. Montgomery's Eighth Army had at last secured what appears to be a conclusive victory against Rommel's forces in Libya and Egypt; Allied forces had hit the beaches in north-west Africa, causing the collapse of Vichy France; the Japanese had been defeated at Midway and were to be defeated again at Guadalcanal; and the USSR had turned the tide against Hitler's forces at Stalingrad. Yet a global grand strategy – an agreed route to defeating the Axis powers – has not yet been agreed. The British and Americans broadly agree on a Germany-first policy (although the US

1. Combined Chiefs of Staff meeting: Marshall (second left) confronts Brooke (third right) and Dill (furthest right)

Navy is dubious about this), as both sides see Hitler's Third Reich as representing the greatest threat.

In November 1942 Franklin Delano Roosevelt, then midway through his unprecedented third term as President of the USA, was considering the possibilities for bringing together the various interests of the great Allied powers into a single consolidated strategy. Roosevelt has since worked hard to forge a good working relationship with Britain and, despite some differences, they are broadly in agreement on the major issues. Finding a way of getting on with Stalin, the Soviet Union's brutal and despotic leader, has been rather more of a challenge; ideologically, they are poles apart.

Roosevelt has tended not to interfere with military decision-making and prefers to focus on grand strategy, diplomacy and politics. Consequently while Alan Brooke (and, before him, you) has had to endure long and exasperating battles with Churchill over specific points of strategy and even tactical decisions, Marshall – Brooke's equivalent in the US Army – has had few such issues. Marshall often did not see Roosevelt for a month or six weeks at a time, but as Brooke has gloomily mentioned, 'I was fortunate if I did not see Winston for six hours!'

That the Allied leaders have arrived in Casablanca for a grand conference is largely Roosevelt's initiative. He had suggested to Churchill back in November 1942 that rather than relying on documents and memos flying back and forth across the Atlantic, it would be useful for the two Allied leaders and supporting staffs to meet face-to-face to thrash out future grand strategy. They had tried to bring in Stalin, but he was too caught up in the ongoing battles around Stalingrad to think of leaving the USSR. Roosevelt and Churchill decided to meet anyway for, as Churchill pointed out, 'we have no plan for 1943 which is on the scale or up to the level of events'. After some discussion they settled on meeting at Casablanca on the Atlantic coast of Morocco in mid-January 1943, with Roosevelt using the codename Admiral Q and Churchill travelling under the codename Mr P.

Now that everyone is here, the differences between the two sides are clear for all to see. The strategic and diplomatic spat at Casablanca could well test Roosevelt's emollient leadership style and Churchill's bullish resolve no end. For the moment it is the Combined Chiefs of Staff who are thrashing out the issues – Roosevelt and Churchill will become involved later. Upon listening to Brooke and then Marshall around the conference table, it is obvious that they differ significantly on how to follow up the successful North African campaign. With the British Eighth Army closing in on Tripoli, and Anglo-American forces sweeping across French North Africa into Tunisia from the west, the view is that by the spring the Axis forces in North Africa will be defeated and the Allied position in the Mediterranean fully secured. But what to do then?

Marshall is clear in his view on what the next step should be. You know, from your time in Washington, that the sixty-two-year-old Marshall is highly regarded as a fine organiser; he is driving forward the vast expansion of the US Army into the war-winning weapon it will need to be in order to defeat Germany and Japan. He is also the key player in providing strategic advice to Roosevelt, even though he has not taken a major battlefield command himself, and it is here that friction is developing between the British and the Americans. Marshall had travelled to Britain in the spring of 1942, armed with what he viewed as the blueprint for Allied victory: the Marshall Memorandum. In part this was based on planning work conducted by General Dwight Eisenhower, who stated that 'We've got to go to Europe and fight, and we've got to quit wasting resources all over the world'; he pressed for a 'land attack as soon as possible' in Western Europe.

Marshall was convinced that the clearest and most logical route to victory against Germany was to concentrate American forces in England in 1942 (Operation Bolero), and then launch a joint cross-Channel attack with the British as soon as possible, preferably in early 1943 (Operation Roundup). This would aid the Russians by drawing German forces to the Western Front, for the last thing the Western Allies want is for the USSR to collapse or for Stalin to seek a negotiated settlement with Hitler. Marshall's Memorandum got short shrift in London; the British regarded it as fanciful and dangerous even to think about a quick return to France, but they agreed to opening the second front in Western Europe in 1943 *if* proper preparations are made – a suitably vague and elastic commitment.

Brooke has informed you that in his view Marshall is 'a big man and a very great gentleman who inspired trust', but equally someone with a 'stunted strategic outlook'. Brooke despairs of Marshall's inability to grasp the issues involved in mounting a cross-Channel invasion and, more importantly, worries about his limited understanding of what would follow and the strategic implications. Brooke is a firm, tough and direct individual, admired and feared by all he commands and respected by those he works alongside. He is quite willing to quarrel with anyone he believes to be in the wrong, including Churchill. Brooke regards the Prime Minister as an inspirational leader, but a strategic loose cannon to be controlled carefully. They have locked horns on many occasions, sometimes to the point where Churchill thought he might have pushed Brooke over the edge. 'Hate him?' Brooke remarked on one occasion about the Prime Minister. 'I love him, but the first time I tell him I agree with him when I don't will be the time to get rid of me.' Brooke is the clear-sighted strategist, a vital foil to Churchill's impulsive and wayward visions, and when it comes to the grand strategic plan for 1943 it is clear when Brooke speaks that he is shaping Britain's approach more than anyone.

Brooke had started to talk down the prospects of a 1943 invasion of France long before Casablanca, but now he ups the ante. He argues that the Allies will be best served by exploiting the opportunities emerging in the Mediterranean, particularly now that the Torch landings in French North Africa (November 1942) have gone well and Montgomery's Eighth Army is closing in on Tripoli. Brooke continues by saying that a leap at Sicily from North Africa might bring down Mussolini's regime and force Germany to commit great resources to holding down the Balkans, currently garrisoned by many thousands of Italian troops.

Marshall glowers and counters that, although there are many strategic advantages to causing Italy's collapse, the difficulty would be that such a campaign could drain resources from the build-up of forces in Britain, massing for the cross-Channel invasion. Extending or expanding the Allies' commitment to the Mediterranean would put at risk the invasion of northern France in 1943, he points out.

Brooke plays his trump card: he announces that the British no longer believe an invasion of France in 1943 is viable. This causes the Americans – Marshall in particular – some puzzlement; Churchill had only recently been in favour of the cross-Channel invasion for 1943, had he not? Churchill, however, has recently shifted his position. The month before Casablanca he finally accepted Brooke's arguments against the cross-Channel attack taking place in 1943. He had previously lambasted Brooke over his Mediterranean strategy: 'you must not think that you can get off with your "sardines" [meaning Sardinia and Sicily] in 1943, no – we must establish a western front and what is more we promised Stalin we would do so when in Moscow.' Brooke had not been part of any such promise, and although the British ambassador to Moscow has pointed out that Stalin might seek a negotiated settlement with Hitler if the Western Allies do not commit to a Western Front in 1943, Brooke has dismissed this as unrealistic. Finally, on 16 December, Churchill acquiesced to Brooke's view that the cross-Channel invasion was not possible in 1943 and that the Mediterranean had to be the focus of Allied efforts.

The British contingent had then flown to Morocco in an uncomfortable converted Liberator bomber, with Churchill, Brooke, Portal, Mountbatten and others all sleeping on the floor of the cabins. Roosevelt had travelled in the utmost secrecy by a circuitous air route via Trinidad and Gambia, eventually arriving at the Anfa Hotel, Casablanca, on 14 January. Churchill and Roosevelt resided in separate villas for the duration of the conference, but as usual their relationship is cordial and friendly.

Marshall and Brooke have respect for each other in many ways, but here their differing views are stark. Over the following few days they continue to debate across the conference table in the Anfa Hotel. Brooke and the British contingent press for the Mediterranean theatre to take precedence in 1943, with a cross-Channel attack into northern France pushed back to May 1944. They also suggest that a cross-Channel invasion before September 1943 is simply not possible anyway, due to the slower build-up rate of forces in England. Even then, Brooke suggests, the size of force available would be insufficient to avoid being bogged down by even limited German resistance. Marshall simply does not agree. The key to a quick victory over the

Germans is to do battle with them in north-west Europe in 1943; he dismisses the Mediterranean as merely nibbling at the edges, and a huge and unnecessary diversion of resources and effort. It would be possible to mount a substantial invasion of Nazi-occupied France in 1943, particularly as the Germans seemed ill-prepared for such an eventuality, but only if Mediterranean operations are closed down and the forces there transferred back to the United Kingdom. Failure to act decisively in 1943, Marshall believes, could lead to American public opinion demanding greater effort in the Pacific against Japan, or even to Stalin despairing of his Allies' 'weakness' and negotiating a deal with Hitler.

What should the Allies do? Roosevelt and Churchill will have to decide which route to take, and you will have to act as intermediary, as you are perhaps the only person trusted by both sides.

Map 1: Competing strategies for 1943 and beyond

AIDE-MEMOIRE

* Brooke has the greater depth of detail to his arguments. His documents and evidence are compelling.

* Marshall's broader view also carries weight. You can't beat Germany in the Mediterranean. Northern France is the one place where the Allies can win the war quickly.

* But the Allies are winning in the Mediterranean. Should they not maintain that momentum?

* What would the strategic implications of knocking Italy out of the war actually be? Does Hitler really care that much about the Mediterranean?

* The Allies will almost certainly have to invade the coast of France at some point. Why delay, when it will have to be the decisive front?

* The cross-Channel attack will be fraught with difficulties and huge risks. Why not take enough time to prepare properly?

* Even if the forces in the Mediterranean are transferred to England in 1943, the Allies will have insufficient forces to win quickly in France.

* Can you risk Stalin doing a deal with Hitler, however unlikely that seems?

THE DECISION

Churchill and Roosevelt must decide who to back.
What do you advise?

▶ To pursue Brooke's Mediterranean strategy and put off
the invasion of north-west Europe until 1944, go to
Section 2 (p. 286).

OR

▶ To follow Marshall's advice and focus on hitting the
beaches in France in 1943, go to **Section 3** (p. 294).

SECTION 2
EISENHOWER'S
MEDITERRANEAN STRATEGY

29 May 1943, Maison Blanche Aerodrome, Algiers

A group of very high-level Allied political and military leaders are milling about at Maison Blanche Aerodrome, surrounded by nervous security personnel. Among the group are Winston Churchill, General George Marshall (US Army Chief of Staff), General Alan Brooke (Chief of the Imperial General Staff) and Hastings Ismay (Churchill's military assistant). Just arriving are Admiral Andrew Cunningham (Royal Navy Commander-in-Chief, Mediterranean) and you, *General Dwight D. Eisenhower* – or Ike, as you are often called.

2. Dwight Eisenhower

Since November 1942 you have been the Supreme Commander Allied Expeditionary Force for the North African Theatre – in effect, the overall chief of Allied forces and affairs in the whole of North Africa and the Mediterranean theatre of operations. It has been quite a rapid rise for you, for as recently as June 1941 you had only been a colonel – and, two years later, you are now a four-star general. Your reputation as a shrewd organiser, analytical planner and careful diplomat brought you to the attention of Marshall, who propelled you to your current position. Since the autumn you have overseen the broadly successful Allied capture of Vichy French North Africa, and you and your staff have begun a not-always-painless integration with the British, who, commanded by Bernard Montgomery, have advanced from Egypt, across Libya and into Tunisia to link up with you. You have already had a couple of run-ins with the awkward Monty, most recently over a bet that he had made with General Walter Bedell Smith, your Chief of Staff, about a B-17 bomber. Monty had claimed that he could capture the town of Sfax in Tunisia by 15 April at the latest; Bedell Smith thought otherwise – 'come, come, General' – and in a moment of levity had casually agreed that if Monty did so, the Americans would provide him with a B-17 Flying Fortress, complete with American air crew, for Monty's personal use. Sfax was duly captured on 10 April, precipitating a telegram to you all in Algiers: 'Have captured Sfax. Send Fortress.' You had been apoplectic.

As you arrive, Churchill is full of bonhomie and pleasantries, and your suspicions grow. You have been briefed by Bedell Smith – recently returned from Washington, scene of the latest Allied conference – that the British are angling for a further expansion of the Mediterranean campaign. You know of Marshall's scepticism about that. But how far will Churchill push it? As you all mill around, Churchill makes a beeline for you. What is he up to?

After Casablanca, where the Americans had been comprehensively outmanoeuvred by General Brooke and the British team over the future of Allied grand strategy, Marshall and his staff had resolved never to get caught out again. For that conference, the British had had one major advantage: they had come properly prepared to argue their case, with a ship anchored close by, packed with administrative and planning staff ready to provide detailed support for the case the British team would make. In contrast, the Americans had been severely undercooked and had had to draw upon your local team, and even the British staff, to work on the American arguments. Brooke turned the discussion in favour of his Mediterranean strategy, much to the irritation of Marshall.

After days of wrangling, Marshall – hamstrung by some internal squabbles, and under a welter of British detail – had been forced to concede to Brooke's grand strategy. Field Marshal Jack Dill, Chief of the British Joint Staff Mission in Washington, who was well liked by the Americans, had managed to broker a deal in which the British gave ground on a few points, but in essence won the day; there would be an invasion of Sicily in 1943, in part because the Americans accepted that the cross-Channel attack could not now be mounted until May 1944. Marshall and the Americans had lost the argument; the American planner General Albert Wedemeyer sardonically conceded that, at Casablanca, 'We came, we saw, we were conquered.' But they resolved never to allow it to happen again.

By 13 May 1943 Tunisia had been captured by the Allies, and the Axis powers had suffered a crushing defeat. When that defeat came, it was calamitous for the Axis powers; some 250,000 troops had been lost, while the campaign in Tunisia as a whole had cost the Luftwaffe a crippling 2,500 aircraft, a figure that Allied intelligence assessors believed was close to half its total fighting strength. It had been a second Stalingrad. The Allies had been planning for the follow-up invasion of Sicily for some months, though not without squabbling. The main battlefield commanders had been focused on winning the Tunisian campaign, leaving your inexperienced planning staff to devise the outline plan for Operation Husky, the codename given to the invasion of Sicily. Montgomery had referred to the first draft plan as a dog's breakfast, and had insisted on significant changes. You do not really see eye-to-eye, although you admire his command skills. You have also heard whispers that Monty has rubbished your military prowess. 'I can deal with anybody except that son of a bitch,' you grumble.*

---

* Monty stated of Eisenhower that his 'knowledge of how to make war, or how to fight battles, is definitely NIL'.

The Sicily invasion plan had been approved at another high-level conference between Roosevelt, Churchill and their teams held in May 1942 in Washington. But the debate at the highest levels has now moved on to what will follow Husky. Marshall is increasingly sure that he was hoodwinked by the British into the Mediterranean strategy and is determined to limit it as much as possible, to focus on the cross-Channel invasion into France in 1944. Yet during the conference, attempts by Marshall to shut down the Mediterranean completely after the capture of Sicily had been blocked, though equally Brooke's grand vision of crippling the Germans by undermining them from the Mediterranean had also been rebuffed.

You had been astonished, however, when you heard that in Washington Brooke had implied that the British were now getting increasingly uneasy about mounting an invasion of France in 1944. Would they ever commit to it? Marshall had threatened to switch the USA's major effort to the Pacific if the British backed out of invading France in 1944. It had been stalemate. When it came to the Mediterranean, the only agreement had been not to agree on anything. Any firm decision had been knocked into the long grass.

Because of that non-decision, you – as the theatre commander in the Mediterranean – have been handed considerable strategic latitude. Without clear instruction to the contrary, it was effectively up to you to run the Mediterranean campaign as you saw fit. Churchill therefore has seen an opening. Within days of the Washington Conference ending, he arrived in Algiers with Brooke to meet you and press the case for a strengthened and expanded Mediterranean campaign. Realising what Churchill is up to, Roosevelt has sent Marshall to shepherd the process and ensure that you do not concede anything dramatic.

This is all in your mind as the respective commanders and leaders exchange pleasantries at Maison Blanche. As you depart from the airport to head by car to Admiral Cunningham's villa for dinner, Churchill intercepts you and insists on travelling with you, much to the irritation of Marshall, who had wanted to brief you first; he is consigned to another car. This could be a politically and strategically dangerous journey for you. Very quickly it becomes apparent that Churchill wants you to follow up Sicily with a rapid invasion of Italy itself. The collapse of

Italy might well open up 'glorious opportunities' to put yet more pressure on Germany through the Mediterranean and the Balkans.

The bombardment continues over dinner. Churchill presses his case three times in three different ways, talking persistently until he has 'worn down the last shred of opposition', as one observer notes. Even Marshall appears to be wilting; he concedes that a small token invasion of Italy to force the collapse of Mussolini's regime may be worthwhile, though he secretly tells you he wants nothing more than that committed to the Mediterranean – no grand visions likely to divert further resources from the cross-Channel attack.

As you listen to the Prime Minister's rhetoric, you have to decide what to do. Marshall wants to concede very little, if anything at all. But the case is not so cut-and-dried, and it is you who will make the decision. 'I could not endure to see a great army stand idle when it might be engaged in striking Italy out of the war,' you accept. But Churchill clearly wants much more. What can you do?

Map 2: Sicily, Italy and then beyond?

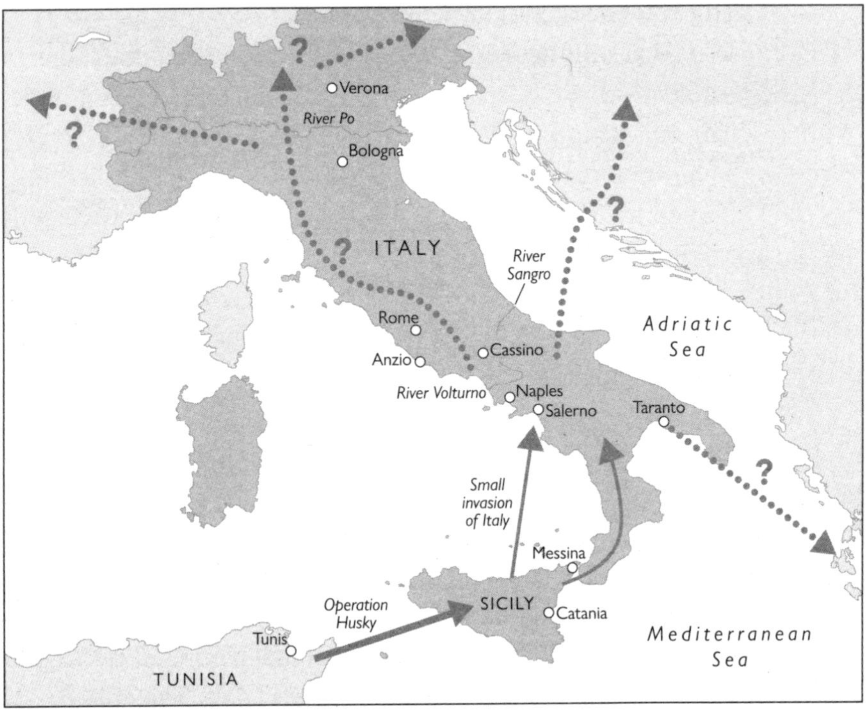

AIDE-MEMOIRE

* After Sicily, a follow-up strike against Italy seems sensible and Marshall is not objecting violently.

* Knocking Italy out of the war would force Germany to send troops to the Balkans (where the Italians currently have lots of forces), and Italy itself to hold back the Allies.

* If all went well, a grander Mediterranean campaign might follow; maybe the cross-Channel invasion might be 'a drop in the bucket' and potentially 'unnecessary'?

* Southern France or even the Balkans become possible follow-up targets, and the latter might even prevent the Soviet Union from 'liberating' all of south-east Europe (examine Map 2 on p. 290).

* As Marshall has previously argued, the Mediterranean could well act as a 'suction pump' and draw in so many resources that the cross-Channel invasion of France in 1944 might be put in jeopardy.

* Liberating France and north-west Europe is the key to winning the war - not the Mediterranean.

* The chances of successfully liberating the Balkans and Greece without using resources intended for the invasion of France in 1944 are almost nil. It will be a massive undertaking and will put back the decisive showdown with the Germans for another year.

* Are the British more concerned about securing their sea route through the Mediterranean via the Suez Canal to India, the jewel in the crown of the Empire? Surely that cannot be anything the anti-imperialist Americans want to be involved in?

THE DECISION

What advice should you offer? You know roughly where Marshall stands on these issues, but this is your command.

▶ To pursue the grand Churchillian vision of an expanded Mediterranean strategy, with possible expansion into southern France and even the Balkans, go to **Section 4** (p. 302).

OR

▶ To follow Marshall's advice and limit any investment in the Mediterranean as much as possible, go to **Section 5** (p. 310).

SECTION 3
THE ROAD TO FRANCE, 1943

16 July 1943, General George Marshall's Allied Expeditionary Force HQ, Bushy Park, England

George Marshall has been putting into place the final pieces of his plan for the Allied invasion of northern France, codenamed Operation Roundup. In just over three weeks amphibious forces spearheaded by the British and American armies will hit the beaches in Normandy to initiate the liberation of Western Europe from the clutches of the Nazis. Since the spring of 1943 planning and preparations have continued apace, although the date of the landings has been repeatedly put back to allow sufficient time to build up the strength required to make Roundup viable. D-Day is now set for 8 August. Intelligence reports have been clear that the Germans are poorly prepared for an invasion and lack fixed defences and troop reserves. Recently, however, reports have been filed that some resources are heading to France, at the expense of the Eastern Front.

Yet although the atmosphere at Bushy Park is electric, all is not well. Concern has been growing over the viability of Roundup, with much of the anxiety emanating from the British. You are Marshall's Chief of Staff, *General Walter Bedell Smith*. Although you have no significant command experience, you have developed a reputation as a fine analyst and a first-rate organiser and staff officer. You had previously worked for Marshall in Washington, before serving as Eisenhower's Chief of Staff in the Mediterranean. When Marshall had been appointed to command Roundup, he had specifically asked for you to return to England to work for him. Over recent weeks you have witnessed the growing scrutiny of Roundup and the intensity of the pressure that Marshall is clearly feeling.

It did not help that Marshall, though he had wanted the command for Roundup, was not everyone's first choice. Churchill had once offered the post to Brooke, but as the Americans are now providing the majority of the troops and equipment, it was clearly politically unacceptable for a Briton to act as Supreme Commander. Eisenhower is still learning the ropes in the Mediterranean, which left George Marshall

3. Walter Bedell Smith

himself to command. Roosevelt had been unhappy at releasing his trusted aide, but there had been no alternative. At least Marshall had the gravitas to enforce his will on other service heads, although it is clear that Brooke thinks him unsuited to commanding an operation of this nature. The British had pressed for General Harold Alexander to be appointed as land commander of the invasion, with Bernard Montgomery to command the British forces and George Patton to lead the Americans. Once these commanders had departed from the Mediterranean in February 1943 to take up the planning of the cross-Channel invasion, it became apparent to all left behind that the Mediterranean had become yesterday's news.

The path to Roundup had appeared at Casablanca a few months earlier, largely as a result of Marshall's lobbying. While General Alan Brooke had argued vigorously for a concentration of resources in the Mediterranean, at least for 1943, the Americans, headed by General George Marshall, had strongly disagreed. The main focus of Marshall's argument had been twofold, and had been all the more compelling because of its inescapable logic. First, failure to commit to north-west Europe in 1943 might have pushed Stalin into a deal with Hitler; the battle raging at Stalingrad had shown that Germany would not now win in the foreseeable future, but it had also demonstrated the enormity of the task confronting the Soviet Union and the scale of the sacrifice it would have to make in order to win. If Stalin had suspected that Washington and London were not willing to pay in blood as well, what would his reaction have been? In addition, if the Soviet Union had turned the tide against Hitler and the Germans had been driven back decisively in 1943, the Red Army would liberate much of Central Europe as well as conquering the Third Reich. Unless the Western Allies were fighting on the continent as well, they would be reduced to the role of bystanders as much of Europe fell under the control of communist rule dictated from Moscow. Either of these extreme options had been possible unless the Western Allies acted in 1943, Marshall argued.

The British had relied on detailed briefing documents which underlined their case that, as things stood, the Allies simply could not muster sufficient force in the United Kingdom to make a cross-Channel attack viable in 1943. Moreover, many preconditions for a successful campaign in north-west Europe, such as assured air superiority and victory in the Battle of the Atlantic against the U-boat menace, had yet to be secured. Marshall countered that many of the difficulties raised by Brooke's team could be overcome by closing down the Mediterranean as soon as possible and transferring resources back to the British Isles. The remainder of the North African campaign could be won with limited resources and Mussolini's regime left to fester.

Brooke had dismissed the idea of Stalin negotiating with Hitler, but a message from Stalin to Roosevelt and Churchill that had emerged during the Casablanca discussions provoked alarm. Stalin had mentioned 'most emphatic warnings' and 'the great danger with which further delay in opening a second front in France' might bring.* The

* The actual words that Stalin used to Churchill in March 1943, when it started to become clear that the Western Allies were going to stall on opening the second front until 1944.

key to the shift in Allied policy at Casablanca had come when Churchill himself began to waver. Brooke had already had to twist his arm that the Mediterranean offered the best option for Britain in 1943, but with Stalin's pointed communication and the Americans pressing for the invasion of France in 1943 to be fixed, Churchill started to hesitate. Brooke had been furious, but could not stem the tide.

As the Casablanca Conference concluded, Washington and London settled on drawing back on any further commitments to the Mediterranean and on focusing resources for the invasion of France in the summer of 1943. The original outline plan for Operation Roundup, sketched out in 1942 by American planning staff, had settled on a combined Allied army of some forty-eight divisions supported by 5,800 aircraft. Landing points had been located between Boulogne and Le Havre. Most of the fighting forces were to be American – thirty out of the forty-eight – but this proved to be an unduly ambitious total, even if troops were withdrawn from the Mediterranean in time for the invasion. Allied planning staff reviewed the existing plan and scaled back the commitment of divisions to thirty-nine and moved the landings to Normandy.

Yet it was still touch and go whether the Americans would be able to train and equip sufficient forces to meet the requirements of any invasion. Herein lay a critical problem, because the Atlantic shipping lanes were dangerous waters indeed in the spring of 1943; the Allies had endured their heaviest losses to the German U-boat fleet in the autumn of 1942 and matters had remained dire early the following year.

Measures to extend air cover across the Atlantic had been developed in the winter of 1942–3, naval escort groups had been expanded and the German U-boat Enigma ciphers, which had been upgraded by the enemy earlier in 1942, had now been broken by Bletchley Park codebreakers. But it was still going to take some considerable time to defeat the U-boat threat once and for all. Will there be sufficient troops and enough materiel in Britain by D-Day to mount the invasion and then press inland?

Another problem identified by Allied planners had been the degree of air cover and support required for the cross-Channel attack. The policy of bombing Germany 'round the clock' – by the American bomber fleet in daylight and RAF Bomber Command at night – had been announced at Casablanca. Yet although it was in part intended to target the Luftwaffe, this had proved much harder to achieve in practice. Although the Germans are not particularly contesting control of the air space over northern France, they still have the resources to return quickly and forcefully to fight for the air space over the English Channel when the invasion begins. There is no obvious way to solve this issue and German air power is, therefore, still a major threat in 1943.

The British, however, have now also started to question the fighting power of the Allied forces, particularly the newly formed units. The Americans have rapidly created and trained new formations, but their tactics and methods have been thrown into confusion after the drubbing that the US Army received at the hands of Rommel in Tunisia at the Battle of Kasserine Pass in February 1943. American units and command systems had been ruthlessly exposed by the wily and battle-hardened Germans, and questions have begun to be asked about the degree to which a rethink is required and whether such a development could be integrated in time for Roundup. Concern is also growing about the technical issues involved in airborne and amphibious operations. The Allies have little experience of such matters, and tactical exercises and training programmes have highlighted many practical difficulties. Integration of air support for land units is also still rudimentary and it is clearly going to be tight, timing-wise, to get everything prepared and ready for D-Day.

In early July, Brooke and the British Chiefs of Staff began raising the pressure by issuing a series of pessimistic and downbeat assessments of Roundup. They have pointed out that the Germans have clearly realised that the Mediterranean theatre has been relegated in importance by the Allies and so have redirected resources to France. The Luftwaffe's growing presence in north-west Europe is being reported by the Allied air forces preparing the ground in France for the invasion, and French resistance units have noted growing anti-invasion measures across the coastal regions of northern France.

What had looked like a risky venture a few months earlier now appears to some people to resemble a gamble fraught with hazard. Brooke has clearly been badgering Churchill and the Combined Chiefs of Staff, the body that provides military-strategic direction for the Allied war effort. Now they too have delivered a downbeat assessment. As you sit and listen to Marshall's woes you begin to wonder whether Roundup might be cancelled. Marshall asks for your views. What advice can you offer?

AIDE-MEMOIRE

* Allied strength is, at best, just up to the task of mounting the invasion, but air control cannot be guaranteed.

* Intelligence reports tell you that the Germans are relatively unprepared in France, a situation that will not persist if the Allies delay into 1944.

* Northern France will ultimately have to be the decisive campaign for the Western Allies. Delaying may cause the war to be extended.

* Do the Allies have sufficient experience to mount a successful, highly complex, multinational, multi-service amphibious / airborne operation?

* Consider the political fallout if the Allies do not open a second front after promising Stalin that they would.

* And what happens after the landings? Do the Allied armies have the tactical know-how and capability to meet the battle-hardened German armed forces in open battle?

* Failure to press the invasion inland from the beaches could well result in a drawn-out attritional campaign, with very heavy loss of Allied life.

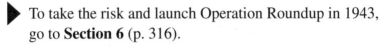

THE DECISION

What should you recommend to Marshall?

▶ To take the risk and launch Operation Roundup in 1943, go to **Section 6** (p. 316).

OR

▶ To postpone the invasion of France in 1944, go to **Section 7** (p. 322).

SECTION 4
ITALIAN GAMBIT

2 July 1943, Italian army headquarters, Rome

An Italian army truck arrives in Rome to deliver a captured British spy to the headquarters of the Royal Italian Army. You are the 'spy', **Dick Mallaby**, a member of Churchill's famous Special Operations Executive (SOE), a group of agents formed to 'set Europe ablaze'. Born in 1919, you lived much of your early life in Tuscany, where your father had inherited an estate. Known as something of a daredevil, you had joined up on the outbreak of war and in 1942 you had been recruited into the SOE. Your Scandinavian looks – blond hair and blue eyes – have resulted in your codename being Olaf. Fluent in Italian, with a strong and genuine Tuscan accent no less, you are also a trained radio operator.

4. Dick Mallaby

You had been flown secretly to Italy a few days earlier to initiate a mission, but had been quickly captured in the aftermath of an air raid, which had raised widespread alarm and had blown your cover.* After being handed over to Italian military intelligence, you fear that the game is up: torture and execution will surely follow. But in a most fortuitous turn of events, the Italians seems to have another purpose for you, one that could determine the outcome of the war in the Mediterranean and may well save your life. It seems that a few days earlier a faction in Italy, headed by Dino Grandi, had been authorised, under the vague direction of the Italian head of state, King Victor Emmanuel III, to explore Italy's future options with the Allies.

5. Dino Grandi

You are told that the King has been alarmed by Mussolini's purge of pro-royalist politicians and officials in the spring of 1943, a measure that Il Duce, the Fascist dictator, thought necessary to clamp down on growing opposition to his leadership. Meanwhile Grandi, one of the senior leaders of the Fascist regime, is more concerned by the deteriorating situation in the war. They have agreed to work together to see what might be done to extricate Italy from the conflict. And they want *you* to be the conduit to Eisenhower in Algeria; you have to convince the Allies that you are who you say you are and that the approach is genuine. As you send your initial message, you include the comment 'Christine, I love you' – a reference to your girlfriend that will hopefully prove your identity. Will the Allied intelligence staff in Algiers believe you and, more to the point, will they act on the offer that the Italian faction is making to open negotiations?

Would the Allies be able to react quickly enough to the possibility

---

* Historically Mallaby was captured in August, but he did play a similar role in opening communications between Rome and the Allies.

of Italy dropping out of the war? Did they have the resources to exploit the situation? Churchill had stayed on in Algiers until early June, thoroughly enjoying himself for a week while Eisenhower fretted over the security implications of the Prime Minister's impromptu visitation. 'I have no more pleasant memories of the war,' Churchill later mused. He was obviously satisfied that Eisenhower, and to a degree Marshall, had been persuaded of the necessity of knocking Italy out of the war. Ike even seemed to be supportive of following an aggressively opportunistic approach to unfolding events in the Mediterranean. Eisenhower, now heading a successful combined military operation across the Mediterranean, did not want to see the momentum of his command lost, and planning for the capture of Sicily continued apace, as did a wide variety of contingencies. Ike in particular saw merit in establishing a grip on the western Mediterranean as a springboard into southern France, even though the British were still obviously lobbying for seizing any opportunities in the Aegean. Much would depend on the speed and effectiveness of Operation Husky – the invasion of Sicily – and on any reaction in Rome. Marshall had departed Algiers deeply suspicious that Churchill had won too much. Time would now tell.

Then Eisenhower was alerted to your message. Was there a realistic option here? Because of your initiative, the Allies seem to think the approach is genuine, although as the messages pass back and forth they seem sceptical about Grandi's ability to deliver on a deal behind Mussolini's back. Grandi despatches a small team to Lisbon to open tentative talks. At Casablanca the Allies had demanded unconditional surrender from the Axis powers, but Grandi hopes there may be a way around this, in the circumstances.* Naturally the Germans will be incensed by any negotiations; there is little doubt that Hitler will come down hard on any possibility of Rome breaking its ties with Berlin. Secrecy is the order of the day therefore. But will the negotiations work out? And can Grandi deliver on his promises?

Eisenhower sends his reliable and astute Chief of Staff, General Walter Bedell Smith, to assess the possibilities. When Churchill and

---

* Unconditional surrender was the formal Allied policy, but historically against both Italy and later Japan they demonstrated some flexibility.

Brooke get wind of the development they become very animated and cable Eisenhower to prepare to seize the moment. Washington is more hesitant, and Marshall warns Roosevelt about getting bounced into further commitments in the Mediterranean, but accedes to Eisenhower's request for extra resources to be made available to exploit any immediate opportunities. The unofficial talks in Lisbon, however, illustrate considerable differences between the Allies and Grandi's faction and they stall for a time.

On 9 July the Allies descend on Sicily and a heavy bombing raid on Rome follows soon afterwards. Now Grandi and his group suddenly feel compelled to act. They urgently and secretly despatch General Giuseppe Castellano and a small team of officers to Malta to meet Bedell Smith again. A key concern for the Italian team is that if Mussolini is replaced, the Germans will quickly move in more troops and seize northern and central Italy. Will the Allies therefore agree to a rapid deployment of troops to bolster the Italians against any German invasion? Eisenhower agrees, but only on condition that Italian soldiers throughout the Mediterranean throw in their lot with the Allies. Grandi consults General Pietro Badoglio, the most senior military commander he can rely on. Badoglio initially wants firmer commitment of support against the Germans, but finally agrees that if Eisenhower can deploy troops quickly into Rome and other key points on the mainland, then he will order the Italian army to switch its support.

Everything depends on tight security and swift action against Mussolini, and only if ordered by the King himself. With the battle in Sicily still ongoing, Grandi's supporters engineer a coup on 25 July in which political control is seized from Mussolini and temporarily handed to the King. Badoglio insists that he take over as temporary Prime Minister, something to which Grandi grudgingly accedes. Mussolini is caught by surprise, arrested and spirited away. Badoglio announces that the Italian armed forces should cease operations against the Allies and calls on the Germans to depart from Italian territory. As he is issuing his proclamation, Allied expeditionary forces begin landing in mainland Italy, Sardinia and Corsica; all are under strict orders to withdraw if the Italians prove hostile. For the most part they do not, and

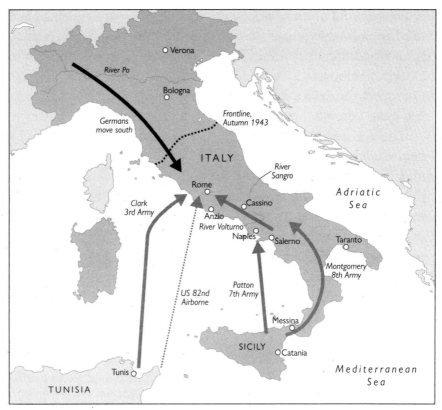

Map 3: Allied Offensives, 1943.

within forty-eight hours the Allies – loosely and sometimes grudgingly supported by bewildered Italian troops – are in possession of many key positions in southern and central Italy, Sardinia and Corsica. German forces deployed in southern Italy and Corsica are pinned down and paralysed, and in some cases overwhelmed. It is a stunning and seismic shift in the Mediterranean campaign.

AFTERMATH

By early August German opposition in Sicily, now isolated and cut off, had come to an end. Monty's Eighth Army was streaming through Messina and racing on up the Italian peninsula, linking up with pro-royalist forces, while Patton's Seventh Army was quickly being shipped to Naples, to prepare itself for a dash to Rome. Badoglio and his government, however, had retreated south from the capital, much to the irritation of the Allies, leaving Rome in chaos. Pro-Mussolini troops held some key positions, supported by some German units, while pro-royalist forces dithered, lacking purposeful leadership.

Hitler had been infuriated by Badoglio's treachery and ordered a number of units based in Austria and southern France to sweep quickly into northern Italy to link up with the few German forces that he had pressured Mussolini into accepting into Italy just weeks earlier. Italian resistance was stronger than expected, and the Germans were repeatedly confounded by local hold-ups and transportation issues. Some infuriated German troops, already enraged by Italian duplicity, resorted to heavy-handed methods and atrocities, causing yet more resistance. The isolated German units further south in Italy were still cut off and hemmed in by Italian units north of Rome, at Foggia and Salerno.

The Allies quickly launched Operation Giant, a predetermined plan to deploy the US 82nd Airborne Division south of Rome at Anzio, with the task of quickly securing key transportation points and linking up with pro-royalist Italian forces.* The port at Anzio was then to be used to allow follow-up forces of the US Fifth Army, commanded by Lieutenant General Mark Clark, to be shipped in for a quick dash to Rome itself. The Allied plans were, however, dependent upon major air support to cover the naval transport of troops, and increasingly the Luftwaffe was appearing in the skies over northern Italy. The build-up of Allied forces at Anzio was too slow and Clark made little headway towards Rome. Patton's Seventh Army, dashing north from Naples along the Italian west coast, finally linked up with Clark's troops and a last push to capture Rome followed, with the city being captured on 17 September 1943.

As autumn fell, Allied forces were advancing north of Rome, while German troops began to dig in along what would be known as the Trasimene defence line south of Florence. It was then that progress stalled. It was considered too late in the year to mount an

---

* A historical plan considered by Eisenhower and his staff.

amphibious invasion of southern France, and in Washington General Marshall was by now fearful of far too many resources being diverted to the Mediterranean – Operation Overlord, the invasion of northern France planned for May 1944, was being put in jeopardy, he argued. Eisenhower, however, began lobbying for a major assault on southern France in May 1944, while in London General Brooke and Churchill were more than happy to support this, as it would keep resources in the Mediterranean, which could then also be used in the Balkans. The British still saw opportunities for troops to be deployed into Yugoslavia and Greece. Brooke's grand strategy of seriously compromising Germany before the cross-Channel invasion was still possible.

THE DECISION

▶ To follow the alternative route go to Section 5 (p. 310).

OR

▶ To explore the history of the Mediterranean campaign and its wider consequences, go to the **Historical Note** on p. 326.

SECTION 5
WAR ON A SHOESTRING: THE MEDITERRANEAN, 1943

9 October 1943, La Marsa, Tunis

A vital conference is taking place between Eisenhower's Allied Force HQ staff and a team from Britain's Middle East Command. Eisenhower himself is present as this meeting, which will finalise the resources and strategy for the next phase of the Allies' war in the Mediterranean. You are **General Henry 'Jumbo' Maitland Wilson**, Commander-in-Chief, Middle East, a post you took up in February 1943, after having served in a variety of command positions in various locations during the Middle East and Mediterranean campaigns. At sixty-two years old, you have a vast wealth of experience to call upon – you had even seen action in the Boer War of 1899–1902. Known as a safe pair of hands, you are nonetheless facing a major challenge in the Aegean, where a rapidly mounted campaign by the limited available British

6. General Henry 'Jumbo' Maitland Wilson

forces is starting to collapse into disaster. You and your team have in part come to see Eisenhower to try and extract further Allied resources to salvage the situation. Churchill has informed you that he has been lobbying Eisenhower to release resources to you, but you are doubtful about what you can get out of him.

The unfolding crisis in the Aegean is in part a consequence of Eisenhower and Marshall's very limited commitment to any wider Mediterranean campaign, a stance they had taken back in May. Despite Churchill's best efforts, he had departed from Algeria with only a loose acceptance from Eisenhower that a follow-up invasion of the Italian mainland was worthwhile, dependent on how the invasion of Sicily unfolded. Beyond that, there was no further commitment.

On the evening of 9 July 1943 Allied forces had begun Operation Husky, their assault on Sicily. It was the first major multinational combined-services operation of the war, involving air, ground and naval elements, principally from the USA, Canada and Britain, all under a unified command structure headed by Eisenhower. For Operation Husky, the Allies had deployed two armies: the US Seventh Army under the abrasive, uncompromising and volatile General George S. Patton, and the British Eighth Army led by the egotistical, forthright and downright rude Montgomery. Although the invasion went well enough, it was not without mishap. Monty and Patton were fierce rivals, and Patton was never going to play second fiddle to the 'little limey fart', as he indelicately nicknamed Monty. Airborne and amphibious forces had endured severe growing pains as many tactical errors occurred, but ultimately Sicily was secured.

The political repercussions of the Allies' success in Sicily were felt most profoundly in Rome. Mussolini's political stock had already fallen dramatically in 1943, and plots were in motion to remove him when he lost a vote of no confidence and was brought down on 25 July and arrested soon afterwards. The new Italian government headed by General Pietro Badoglio had been paralysed by indecision, however, and no attempt was initially made to contact the Allies to extricate Italy from the war. Equally nothing was done to prevent the Germans, under orders from an infuriated Hitler, marching tens of thousands of troops into northern and central Italy and seizing control. The

Germans soon established a new puppet regime – the Italian Social Republic – eventually with Mussolini, whom they rescued in a daring special-forces raid in September, as head of state.

Eisenhower attempted to exploit the situation as fully as possible. His commanders were soon planning a series of landings and operations against mainland Italy, to be initiated when Sicily was fully secured. Monty's Eighth Army struck across the Straits of Messina at the toe of Italy and sent forces into Taranto, while Lieutenant General Mark Clark's US Fifth Army landed at Salerno, close to the important port of Naples. Eisenhower was by then also being made aware of the first peace feelers from Badoglio's new Italian government. If the new Italian regime could be brought onside against the Germans, then rapid and striking success across southern Italy might be achieved.

Eisenhower only had limited resources available, however, as the Allied war effort was already starting to focus on preparations for the cross-Channel invasion – or at least this was what Marshall was enforcing, as part of his strategy of preventing the British from dragging the US into further operations in the Mediterranean. This policy was underscored at the Quebec Conference in August. Marshall, now backed by all his American senior officers and commanders, had demanded that all major Allied resources in Europe be firmly committed to the invasion of France – by now renamed Operation Overlord and pencilled in for May 1944.

By the time Allied troops landed on the Italian mainland the theatre had been relegated to a decidedly secondary status; Washington would only countenance a low-level commitment to Italy, with a further aim of capturing Corsica and Sardinia to enable a secondary invasion of southern France to support Overlord the following year. This was all to be accomplished on the cheap; the Americans had even withdrawn resources from the Mediterranean to stop the British using them elsewhere in the Mediterranean. They still suspected that Churchill had designs on the Greek islands, particularly Rhodes, an outline plan for which, called Operation Accolade, had been agreed in principle at Casablanca. But now that the Americans had focused their resources elsewhere, Accolade had been shelved – that was unless the British decided to act alone.

The limits on independent British power were soon becoming apparent, however. When Italy surrendered on 8 September, you had been instructed by Churchill to act quickly, with whatever you had, to try and seize key islands and bases in the Greek islands – in effect, a scaled-down Operation Accolade. Churchill believes that securing them would ultimately prevent the Soviet Union controlling Greece in the post-war world. Despite your having only very limited resources, especially to mount an offensive operation, Churchill ordered you to act before the Germans could take control of the Greek islands from the Italians. You had been somewhat sceptical about the operation, a view reinforced by rumours that even General Alan Brooke, the Chief of the Imperial General Staff, is opposed to it.

Initially, when there had been only limited opposition, the Aegean campaign had gone well, but soon the Germans had asserted their influence and your forces had started to come under intense pressure. By early October the whole scheme was starting to unravel. The island of Kos has just been lost, along with almost 1,500 British soldiers; the destroyer HMS *Panther* has been sunk; and the crucial island of Leros appears to be the next likely target for the Germans.*

As you and your team discuss the matter with Eisenhower and his staff, it becomes patently clear that no help is available. Ike point-blank refuses. It is contrary to agreed Allied policy to open up more fronts in the Mediterranean, and in any case, Eisenhower explains, there are going to be few enough resources to fight the Italian campaign as it is, without committing elsewhere. Under direction from the Combined Chiefs of Staff, the great majority of resources are now earmarked for Overlord, and Marshall is not going to ease up on that.

AFTERMATH

The British eventually had to abandon their plans for the Aegean and withdraw what remained of their forces deployed there, but with quite

---

* One of these actions formed the basis of the classic war movie *The Guns of Navarone* (1961).

significant losses. More than 5,000 troops were killed and wounded, while close on 45,000 were lost to captivity. It proved to be Germany's last significant success of the war.

Eisenhower's prognosis on the available resources for the Mediterranean was accurate. The Italian campaign soon became bogged down, and the Allies' lack of commitment tempted the Germans into committing to a forward defence of the Italian peninsula, forcing the Allies to slug their way north. It degenerated into an attritional strategic sideshow, in which the Allies were able to pin down German resources without it ever proving that decisive. For Churchill, his hopes of retaining control of Greece to keep the Soviet Union out of the eastern Mediterranean had to be pinned on doing side-deals with Stalin, much to the irritation of Washington.

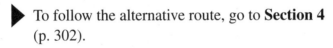

THE DECISION

▶ To follow the alternative route, go to **Section 4** (p. 302).

OR

▶ To explore the history of the Mediterranean campaign and its wider consequences, go to the **Historical Note** on p. 326.

SECTION 6
OPERATION ROUNDUP, 1943

8 August 1943, the beaches of Normandy

It is first light on the coast of Normandy, northern France. Overhead, Allied aircraft patrol the skies, attacking the Germans' defensive positions, while offshore a heavy naval bombardment is in full sway. The Axis powers' coastal troops cower in their inadequate fortifications and trenches. D-Day has come and the Allies are embarking on their much-vaunted liberation of Western Europe – Operation Roundup.

Back in England, the leaders of the forces now landing in Normandy nervously wait for news. Although the Supreme Commander of the Allied Expeditionary Force in Roundup is General George Marshall, the landings are commanded by you, *General Harold Alexander*. Of aristocratic bearing, schooled at Harrow and trained at Sandhurst, you have been provided with all privileges and skills necessary for a high-profile military career. A well-regarded performance in the First World War, calm leadership in the debacle of the British Army's performance in 1940 and then stints commanding in South-East Asia and North Africa have brought you the command of the greatest amphibious invasion in history, Operation Roundup. It is a massive responsibility, but you have relied heavily on your subordinate battlefield commanders: the charismatic and dynamic George Patton and the authoritative and uncompromising Bernard Montgomery. Patton's First US Army is landing at the foot of the Cotentin peninsula, while Montgomery's Second British Army is arriving in the Bayeux / Caen sector. Glider and paratrooper forces of the US 101st and the British 6th Airborne Divisions have been dropped and deployed to secure the flanks of the landing beaches and to capture key positions. Overhead the Allies have mustered considerable air strength, while offshore in the English Channel their navies naturally dominate.

The D-Day landings go well enough for the Allies, to the great relief of yourself and the other senior commanders – Brooke, in particular, had been fretting about your chances. Just a few weeks ago Marshall had informed you that Brooke even tried to get Roundup cancelled, but the Combined Chiefs of Staff, and Roosevelt himself, had stepped in. Extra resources were assigned to the invasion, par-

7. General Harold Alexander

ticularly some more experienced follow-up troops to be transferred from the Mediterranean.

Curiously, the response to the Normandy invasion from the enemy appears initially sluggish. Hitler and his staff are in the middle of coordinating the Axis response to the Soviet counter-offensive around Kursk; they may be distracted? Your intelligence staff have informed you that Rommel has recently been sent to command the defence of France, but that was only a few weeks ago, far too late to make any material difference. All your other intelligence sources have continued to highlight the weak defences of the Germans in France.

Yet not everything has gone well. Chaos initially reigns on the landing beaches, and although German opposition is slight, the confusion – along with the organisational hurdles that the assault forces have to overcome – proves enough to slow the Allies down. The airborne landings have been only moderately successful at best, and the Allied air forces are repeatedly challenged in the skies over the English Channel. German fighters have recently been freed from defending German cities against the Allied bombing campaign, as the

RAF's Bomber Command and the American strategic air forces have switched their attention to attacking the continental transport system to hinder the enemy's troop movements. In the first few days of the landings, despite being heavily outnumbered, the Luftwaffe contests control of the air space over Normandy with some determination. The RAF and US Army Air Force pilots and air crew resist desperately, attempting to keep the German air attacks at bay, but heavy losses are suffered and the landings are put under considerable pressure.

The Allies establish themselves in Normandy, but progress soon slows. As you oversee operations, you begin to realise some of the problems that will have to be confronted. Monty's forces start to become embroiled in a series of attritional battles between Caen and Falaise, while Patton's forces, after struggling through difficult and dense terrain, soon find themselves confronted by fresh German units rushed into northern France from the Mediterranean. With the Balkans still garrisoned predominantly by Italian troops and the Allied threat to Sicily and southern Italy receding, German forces are freed to redeploy to France and although they are slowed by Allied air attacks, it is not enough to prevent them arriving in time to play a role.

Allied ground units – mostly inexperienced and new to battle, and unsure of battle techniques and tactics – find themselves bogged down against a skilful and resourceful enemy. In contrast, the Germans appear to be unable to mass sufficient strength to mount their own serious counter-attacks and their losses seem to be growing steadily. In late August you receive a report that Rommel has been killed in an air attack, causing Berlin to transfer Field Marshal Erich von Manstein, fresh from stunning successes on the Eastern Front, to France to coordinate a counter-attack against the Allies, but as autumn closes in, the battle is becoming static.

As the autumn rains hinder mobility and manoeuvre, progress in the campaign in France slowly but surely bogs down. Monty manages a series of careful and orderly offensives that see his forces across the Seine, capturing Rouen in early October. Then he halts to clear the port of Le Havre, to aid reinforcement and supply. Patton's forces have closed in on Paris, but are then embroiled in a series of battles of

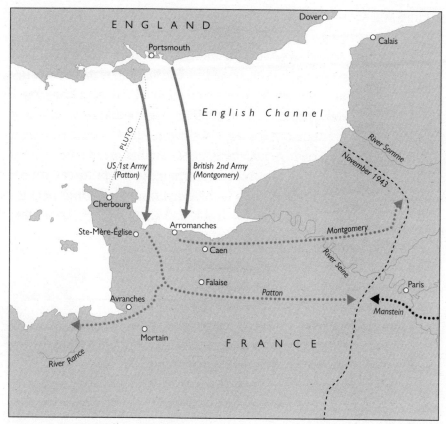

Map 4: Operation Roundup, 1943

manoeuvre with German troops commanded by Manstein. It is a titanic battle of wills between two of the most dynamic and aggressive operational commanders of the war, but by the November they have battled each other to a standstill.

AFTERMATH

As winter sets in, the Allies can reflect that they have at least established themselves in France, taking pressure off the Soviet armies in the east. Yet progress in the west has been limited. By the end of 1943 the Germans have effected a defensive line that will take some considerable effort to rupture. It did not help that Mussolini's recovering regime was now feeding troops into France to bolster the German effort. The campaign in 1944 will be no easy advance.

As you and Marshall review Roundup, you can at best regard it as a modest success. Allied forces are at last battling the principal enemy on the continent, and while you have enjoyed no blitzkrieg-style breakthrough, your forces have nonetheless learned a great deal and established the groundwork for the 1944 campaign. London, however, is dismayed at the growing casualty lists; the British War Office knows that its ability to supply reinforcements to replace the losses on the continent is strictly limited, such is the drain on their pool of personnel. Much still has to be done in 1944, and the prospects of a long, drawn-out, attritional campaign loom large.

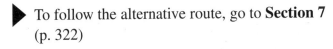

THE DECISION

▶ To follow the alternative route, go to **Section 7** (p. 322)

OR

▶ To explore the history of the Mediterranean campaign and its wider consequences, go to the **Historical Note** on p. 326.

SECTION 7
D-DAY DELAYED

23 July 1943, Combined Chiefs of Staff Building, Constitution Avenue, Washington

General Alan Brooke, the British Chief of the Imperial General Staff, has just begun his briefing to this emergency meeting of the Combined Chiefs of Staff (CCS), the body that advises the senior Allied leadership on the military and strategic conduct of the war. The meeting is tense, as the future of Operation Roundup – the Allies' proposed invasion of northern France, now just a few weeks away – is being debated. As Brooke outlines all the concerns about Roundup from British and American officers based in the United Kingdom, he finds that he is pushing at an open door. The meeting proves to be seismic in its implications.

You are **Field Marshal John (Jack) Dill**, Britain's senior representative on the CCS; in effect, the voice of Churchill in Washington. Your relations with the Americans are excellent, so when Brooke and his team flew over from the British Isles yesterday you were able to brief them on the state of play and the prevailing attitudes to Roundup. Brooke needed to know the feelings in the US capital. Are the Americans still hard set on Roundup, or have doubts filtered across the Atlantic, too? General George Marshall, the Allied Supreme Commander for Roundup, is now based in the south of England, putting together the finishing touches to his plan. He has steadfastly batted away all the growing concerns about the invasion, posed by American as well as British commanders and planners closely involved with the operation.

As a way of dampening down the speculation, it had been he who suggested that Brooke take the matter to the Combined Chiefs of Staff in Washington, never expecting that the British would do exactly that. Now Brooke is in Washington, having called Marshall's bluff.

The key figure in the meeting is Marshall's replacement as the US Army's Chief of Staff, General Douglas MacArthur, a senior figure in the US military who had previous experience of the post in the 1930s. MacArthur is a controversial, high-profile figure in the USA, who had suffered a heavy defeat in the Philippines at the hands of the Japanese in 1942 and had, prior to his appointment as Marshall's replacement as Chief of Staff, been desperately trying to build an alliance to support his strategy of seeking revenge. Consequently he was not best pleased when the President called him home; he would much have preferred to retain his field command in the Pacific theatre. Roosevelt, however, insisted that he needed a leader of great stature to act as his senior military advisor and MacArthur grudgingly returned to Washington.

8. Douglas MacArthur

MacArthur is well aware of the negative whispering campaign against Roundup, and you suspect he would have been quite happy to send Brooke away with a flea in his ear. Yet Admiral King, the US Navy's Commander-in-Chief, sees this as a real opportunity to claw back resources for the Pacific War. King and the navy have been forced to take a back seat for 1943 as the focus of American power has been centred squarely on Europe and on Marshall's Roundup plan. King had wanted greater commitment to the Pacific and to take the war to the Japanese, especially in the wake of American success at Guadalcanal, but at Casablanca this had been denied. King is cantankerous and confrontational, not well liked (even if respected), but as you look on, he seizes his moment to wrest control of Allied grand strategy back from the 'Europe first' school of thought, even if this means agreeing with the British – something the Anglophobe King always finds difficult. MacArthur, too, is aware of the possible outcome of Roundup being ditched, most obviously that resources might be freed for Pacific operations, which are still close to his heart.

You had briefed Brooke that you believe King and MacArthur might be willing to drop Roundup if the negative opinions are convincing enough. Brooke may have to be careful, you warn, that the Americans might refocus their energies away from Europe entirely; Brooke certainly does not want that. But a deal is quickly cooked up that sees Roundup cancelled until May 1944 to allow for proper preparations, alongside a reallocation of resources for 1943 and early 1944. King and MacArthur will switch their focus to the Pacific for the next few months; the American army air forces in Europe will concentrate once more on the bombing campaign against Germany (much to their relief); while Brooke reactivates plans for the invasion of Sicily, now pencilled in for September 1943. These recommendations are passed on to a surprised Roosevelt and a relieved Churchill, who sign off on them. Marshall is apoplectic when he discovers how he has been outmanoeuvred by King and Brooke, and even considers resigning his command in protest. The smooth and emollient Harold Alexander, the land-force commander for Roundup, and the sharp and astute Air Marshal Arthur Tedder, Marshall's deputy Supreme Commander, eventually talk him round. Marshall decides to stay on and command Roundup (soon to be renamed Overlord) in May 1944).

AFTERMATH

The political and diplomatic fallout of the CCS decision is a more pressing matter, however, for Roosevelt and Churchill. How can they persuade Stalin that he has not been abandoned by the Western Allies? It does not help that just as the notification that Roundup has been postponed until 1944 reaches Moscow, the Germans launch their Kursk offensive, which brings about the largest tank battle in history and eventually costs the Soviet Union close on 200,000 casualties – even before they launch their own counter-offensives, which, though strategically successful, result in more than half a million more losses.

As the maelstrom flares around Kursk, Stalin issues a stream of insulting and dismissive messages to his 'Allies', decrying their weakness and highlighting their obvious desire to win the war with Russian lives. Churchill's attempt to prove commitment by highlighting the proposed invasion of Sicily is dismissed, and even the redoubled bombing campaign against German cities – previously heartily endorsed by Stalin – is met with indifference. Panic breaks out in London and Washington in early August when intelligence sources in Stockholm become aware of secret meetings having taken place in Sweden between German and Soviet representatives. It is always a concern for Churchill and Roosevelt that Stalin, if he felt the Soviet Union could get a reasonable deal out of Berlin, might negotiate a separate ceasefire or come to some arrangement with Hitler; this was not unprecedented, after all, as the Molotov–Ribbentrop agreement of August 1939 had demonstrated.

Churchill, Brooke and Roosevelt's long-serving Secretary of State, Cordell Hull, fly to Moscow in August armed with concessions and promises of more material aid, alongside a cast-iron commitment to the invasion of north-west Europe in May 1944. They are met with a frosty reception, but in the end Stalin is obviously bluffing. He is buoyed by the success, however costly, of the fighting around Kursk; the Germans have given it their best shot and have been thrown back, with crippling losses. Stalin can now see a path to victory and the meetings in Sweden were little more than a feint intended to wring more support from the Western Allies. The Grand Alliance survives, just.

▶ To follow the alternative route go to Section 6 (p. 316)

OR

▶ To explore the history of the Mediterranean campaign and its wider consequences go to the **Historical Note** on p. 326

HISTORICAL NOTE

The historical route through this chapter is to back the British vision at Casablanca (from Section 1, go to Section 2). Brooke and his well-prepared and fully supported staff comprehensively out-argued the Americans and secured their strategy for an invasion of Sicily, and possibly more, for 1943. Marshall, however, did not allow himself to be outmanoeuvred again, and in the various debates and conferences later in 1943 he blocked British hopes for a major campaign against Italy and the Balkans (from Section 2, go to Section 5). He was per-suaded by Brooke and Eisenhower that a small token invasion of Italy would be useful, but no more. This limited campaign tempted the Germans into thinking that Italy could be held, and so the campaign there dragged on until the closing stages of the war, without achieving a great deal, save for tying down German troops and resources.

But what if Eisenhower had been persuaded to lobby Washington for more resources in mid-1943 and Marshall had acquiesced (from Section 2, go to Section 4)? If coupled with a more amenable faction in Rome, one willing to cooperate quickly, Eisenhower might have had the resources to press a much more energetic campaign (the Allies had a whole range of contingency plans for such an eventuality) and deliver a knockout blow to the Germans. Berlin was already hesitating about committing to the forward defence of Italy, so it is possible that, with rapid, vigorous actions, the Allies might have been operating north of Rome in 1943, a year ahead of schedule. Yet even if this had occurred, Marshall's reservations about how important it would prove might well have played out. It would have been difficult indeed for the Allies to exploit the Mediterranean any further, although it would have offered a useful springboard into the South of France, which would have taken the pressure off the cross-Channel invasion, although this would still have been necessary. The expanded Mediterranean campaign would also probably have favoured Britain's hopes for rescuing the southern Balkans from the clutches of the advancing Red Army, but without American support such ventures would always have been risky.

If Marshall had prevailed at Casablanca, would Roundup have gone ahead (from Section 1, go to Section 3)? With no commitment to a Mediterranean campaign, and with a careful massing of resources, it would technically have been possible, but some of the preconditions for historical success in Overlord in 1944 would not have been met. The Allies would almost certainly not have won air supremacy (this was a product of the bombing offensive, and was not achieved until the spring of 1944), nor would they have learned all the lessons of the amphibious and airborne campaigns that were picked up historically in the Mediterranean in 1943. Their forces would also have been less battle-hardened and more inexperienced if they had not gone through their paces in the Mediterranean. The outcome of pressing ahead with Roundup (from Section 3, go to Section 6) would most likely have been a stalemate and a much longer and attritional land campaign for the Allies in Western Europe. Yet it might well have aided the Soviet war effort by drawing resources away from the Eastern Front, and so could in certain circumstances have brought forward the end of the war, if at a significant cost. It might also be questionable whether aiding Stalin more was a positive outcome for Churchill and Roosevelt.

Could Roundup have been pulled at the final hour (from Section 3, go to Section 7)? Such an outcome was unlikely once it had got so far, but a coalition of Brooke and MacArthur, with King lurking in the background, might – if backed by the right evidence (of which there would have been much) – have colluded to bring Marshall's plan down. Brooke was always fiercely opposed to Roundup; he was even wary about Overlord in 1944.

Ultimately the argument between the British and Americans, and principally between Alan Brooke and George Marshall, was one that shaped the final outcome of the Second World War. The decisions they made set the path to the post-1945 world and were in part responsible for the development of the Cold War and the Iron Curtain. It is doubtful if alternative choices would have improved matters, and of course this was a global war; there was still the matter of Japan to settle, and so prolonging the war in Europe was never palatable.

Jack Dill served with great distinction in positions of high political and strategic influence until his untimely death from aplastic anaemia in November 1944. He was regarded with great affinity in the USA and was buried at Arlington Cemetery.

Though he never commanded operationally, **Dwight Eisenhower** went on to head the Allied forces in Western Europe through the 1944–5 campaign, earning many plaudits for his diplomatic and considered leadership. He eventually went on to serve two terms as President of the USA.

Walter Bedell Smith served as Eisenhower's Chief of Staff from 1942 until the war's end, earning the reputation of Ike's hatchet man. Cool and analytical, he was pivotal to Allied command. In the post-war world he served as Director of the CIA before working for Eisenhower in a series of posts.

Dick Mallaby (full name Cecil Richard Dallimore-Mallaby) was in reality captured later than in the scenario given here, but did fulfil a vital role in communicating between the new Italian leadership and the Allies in mid-1943. He continued to work for the SOE until the end of the war. Afterwards he married Christine (of the 'Christine, I love you' message) and worked for NATO.

Henry 'Jumbo' Maitland Wilson performed a variety of important Allied command roles without ever gaining the accolades of some of his more illustrious colleagues. Well liked and respected, he retired as Field Marshal and later served as Constable of the Tower of London.

Harold Alexander, later Earl Alexander of Tunis, eventually became Supreme Commander of the Mediterranean. Much liked by Churchill, he was mentioned as a commander for Overlord and then as a replacement for Monty. He later served as Governor General of Canada, and later still as a minister in Churchill's government of 1951–5.

7

RACE TO THE RHINE:
ARNHEM AND OPERATION MARKET GARDEN, SEPTEMBER 1944

SECTION 1
EISENHOWER'S DECISION

10 September 1944, Melsbroek Airport, Brussels

For the first time in weeks the Supreme Commander of the Allied Expeditionary Force in Western Europe, General Dwight Eisenhower, is about to meet one of his primary field commanders, Field Marshal Bernard Montgomery. 'Ike', as Eisenhower is often known, has flown from his base in France to Brussels, largely at Monty's insistence, to discuss Allied strategy, thrash out a plan for 21st Army Group (Monty's command) and, hopefully, put an end to the brewing tensions and fractiousness that have started to creep into relations between Monty and Ike. Eisenhower is not best pleased at having to fly to Brussels, partly because he is labouring with a knee injury suffered when helping to manoeuvre an aircraft on the ground a few days earlier. He is incapacitated to the point that he will not be able to leave his C-47 Skytrain transport aircraft, and so this vital meeting – one that could decide the next phase of the Allies' advance towards victory over Hitler's Germany – will have to take place on board the aeroplane.

Ike has travelled with Lieutenant General Humfrey Gale (his logistics chief) and you, **Air Marshal Arthur Tedder**, Eisenhower's Deputy Supreme Commander. With a military career dating back to the Great War, you are regarded by your peers as a keen intellect with a sharp, analytical mind. You studied history at Cambridge before joining the army in 1914, from where you eventually advanced into the Royal Flying Corps and then the RAF. After steadily advancing through the interwar years, your major break came when you were appointed to head the RAF's Middle Eastern air forces in 1941. When Monty arrived in the desert in 1942 you initially got on well with him, but as you both advanced across the Mediterranean and then returned to the UK to lead the Normandy D-Day invasion in 1944, your relationship has soured. Your faith in Monty has dissipated, after what you regarded as a lacklustre and unimaginative performance in Normandy – you even lobbied Ike to have Monty replaced. Now you and Eisenhower are having to fly to meet Monty, rather than the reverse, which would have seemed more appropriate – you think Ike is very long-suffering when it comes to dealing with Monty.

1. Arthur Tedder

Both you and Ike are fully and ruefully aware that Montgomery is a very tricky customer to deal with; caustically rude about many of his contemporaries, infuriatingly arrogant and increasingly convinced of his own genius and infallibility, he has won few friends. Even those who admire Monty's undoubted professionalism and military skills find it hard to like a man who believes describing a fellow commander as 'a good plain cook' to be a compliment, and who wooed a prospective girlfriend with his thoughts on tank tactics (unsuccessfully, it must be said).

Yet Monty is the British national hero of the hour, the military leader who had saved the day in the desert war against the great German Field Marshal Rommel at the Battle of El Alamein in 1942, and who has masterminded a series of great victories ever since, culminating in his triumphant command of the Allied forces on D-Day in June 1944. Following his great victory in Normandy, Monty has been agitating to seize the strategic initiative once again, something made more difficult since Eisenhower has now taken over command of the campaign, as per pre-invasion planning. Clearly Monty could not continue to exert overall control of the Allied forces in Europe when increasingly the majority of them were American. Yet he still continues to badger Ike and his staff to provide extra resources for Monty's 21st Army Group, now advancing rapidly into the Low Countries, and to back his vision for the next stage of the campaign. He has been agitating for this meeting with Ike for some time, not having met him in person since 26 August, and then only briefly.

As the Skytrain lands, you and Ike both know that it will probably be a difficult meeting with Monty, but Ike is the consummate communicator, well versed in dealing with highly strung, egocentric personalities whilst bringing out the best in his awkward charges. In your view, Ike may lack the military grip and experience of commanders such as Monty, but he brings considered political oversight and vision to his role as Supreme Commander of the Allied forces in Europe. Ike's overriding concern is balancing the conflicting visions of his senior generals for the final push towards Germany, in particular Monty, who has been railing against Ike's preferred grand strategy in an increasingly tetchy manner.

Ever since German resistance in France crumbled in mid-August, leading to the liberation of Paris on 25 August, the Allies have been racing towards Germany at full speed, pursuing their routed enemy. The German collapse has been sudden and cataclysmic. Of two whole armies that fought in Normandy, almost half a million troops have become casualties or been captured by the Allies, whilst out of seven whole armoured divisions, barely 1,300 soldiers and a few tens of tanks have escaped across the Seine to flee back to Germany.

By early September Canadian forces have arrived in Brussels, British troops have seized intact the vital port of Antwerp, and American units, particularly those of the flamboyant and dynamic General George Patton's Third Army, have raced to the very frontier of Hitler's Third Reich itself. This stunning success has, however, brought huge supply problems: fuel, food and ammunition are still largely arriving in Normandy, now hundreds of miles behind the front lines, and are having to be transported across France, often by road as the rail network is still in tatters. Quite simply, there are not enough supplies to feed all the forces racing towards Germany. Eisenhower now faces key choices – choices that might, if all goes well, bring the war to an end in the autumn of 1944, but that will certainly result in a far easier push into Germany itself, when the time comes.

On 23 August Eisenhower had decided upon a two-pronged advance towards Germany in which Montgomery's 21st Army Group (made up mainly of British and Canadian forces) headed towards the Ruhr via Belgium, while General Omar Bradley's 12th US Army Group pushed eastwards towards the Saar; both the Saar and the Ruhr were vital to Germany's lingering war effort.

However, Monty had badgered him into giving greater emphasis to his advance in the north, including the use of American troops. Bradley and Patton had been furious and continued to push forces towards Metz, despite dwindling fuel supplies, flagrantly in contravention of Ike's plans. Bradley fumed, 'we've got to make it clear to the American public that we are no longer under the control of Monty.' Back in Washington, Eisenhower's boss, General George Marshall, has also been irked; he is coming under increasing media and political pressure to ensure that US forces are seen to be playing the dominant role. He has instructed Ike that the British must not be seen to be leading the way to victory when the majority of Allied troops deployed in Western Europe are American; Eisenhower is therefore to get a grip on the political situation. And yet despite the emphasis being placed on the northern route, Monty is still grumbling; he wants more.

Monty clambers onto Ike's aeroplane accompanied by his chief administrative (supply) officer, Major General Miles Graham. Both Monty and Graham eye Ike's supply officer, Humfrey Gale, with suspicion – both think him 'quite useless'. Monty sets the tone immediately by asking for Gale to be removed from the meeting. He then launches into a hectoring lecture about the inadequacies and failings of Allied strategy. He waves under his superior's nose the recent telegrams he has received outlining Eisenhower's strategy, asking, 'Did you send me these?' Eisenhower confirms that he did, to which Monty barks, 'Well, they're nothing but balls, sheer balls, rubbish!' Eisenhower, who learned in his younger days to control his temper, defuses the situation by patting the Field Marshal on the knee, gently stating, 'Monty, you can't talk to me like that. I'm your boss.' Montgomery apologises and eases off a little, but the meeting continues in a

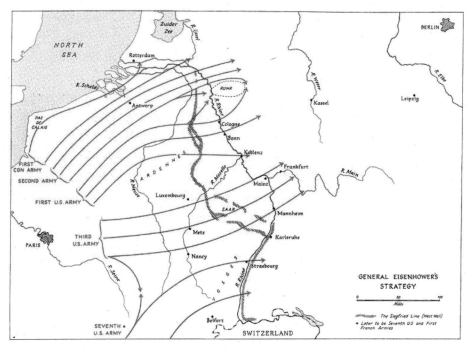

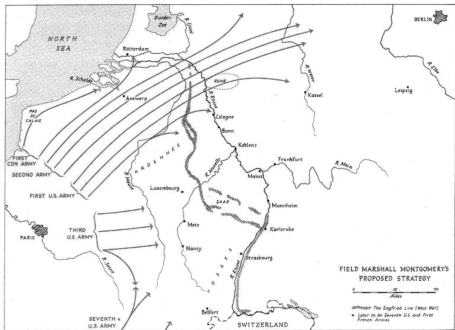

Map 1: A broad front versus a single thrust

disagreeable manner. Miles Graham regards it as a 'complete dog-fight', whilst Gale thinks that Monty is vague, unsure of his facts and trails 'a number of red herrings across the meeting'.*

The crux of the disagreement centres on the thrust – or lack, as Monty claims – of Allied strategy. Monty is convinced that if Ike backs him with all that he needs, he will be able to secure a crossing over the Rhine and launch a concerted strike into northern Germany, potentially delivering a knockout blow to the reeling Germans. Monty's forces have sorted themselves out preparatory to the next thrust, and he still believes the opportunity is there to finish the job. With Germany's allies peeling away from Hitler's regime – Bulgaria has just changed sides, whilst Finland and Romania both look likely to drop out of the war in the next few days – a push across the Rhine could be all that is needed to topple the Third Reich. Not to press this single 'full-blooded thrust' threatens to prolong the war, Monty argues.

Ike disagrees, labelling Monty's plan 'nuts'. 'You can't do it,' he argues, and states that a single thrust down one axis of advance is not feasible. Monty would have to keep dropping off divisions to guard the flanks, denuding the spearhead of what it needs in order to maintain any real momentum, Ike points out. Miles Graham assures you that even without the immediate use of the major port of Antwerp (still not open), he can supply a thrust right across the Rhine and deep into Germany. You and Ike are deeply sceptical. Furthermore, there is the political dimension: the only way Monty's plan can work is if Ike closes down all offensive operations everywhere else and throws everything behind the forces commanded by Montgomery. The notion that Bradley and Patton can be completely halted, in order to propel Monty's forces deep into Germany, is clearly unacceptable politically, even if it does make military sense, which in Ike's eyes (and in your view, too) it does not. Eisenhower, backed by you, holds his line and states that a wider campaign against German forces will be maintained, although the northern push, composed of a mixed force of British, Canadian and American troops under Montgomery's leadership, will still receive greater backing than other forces. The Field Marshal is frustrated.

---

* Historically it is unclear whether Humfrey Gale left all or part of the meeting.

And then Monty plays his wild card. Even if Ike will not back him wholeheartedly on a push into Germany, he now offers a way to at least seize a crossing over the Rhine: Operation Market Garden. It is an audacious plan, which calls for Allied airborne forces – 60,000 American, British and Polish troops – to be dropped by parachute or to land in gliders up to sixty miles behind the enemy's lines to secure a corridor along which British ground forces, spearheaded by the 20,000 vehicles of XXX Corps, would advance at breakneck speed over a series of bridges at Eindhoven and Nijmegen, all the way to the Lower Rhine at Arnhem in the Netherlands. The bridges are the key to

Map 2: Operation Market Garden plan

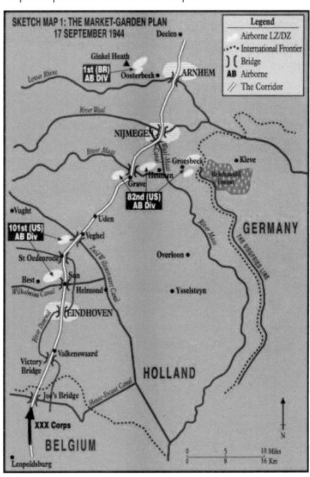

it all and would have to be seized 'with thunderclap surprise' at the onset of the operation to prevent their destruction by the Germans. After the capture of the bridge at Arnhem, further operations might flow, but a crossing over the Rhine would in any circumstances be a major strategic achievement. To say that the plan is bold would be a serious understatement, and it is quite out of character for the usually cautious and risk-averse Monty to propose anything like it. You are startled by the audacity of it all.

To make the plan work, Montgomery requires the deployment of the First Allied Airborne Army, commanded by the American Lieutenant General Lewis Brereton. Established just a few weeks before, the First Allied Airborne Army comprises all of the Allies' airborne forces and is sitting in England eagerly waiting to be used. You know that Eisenhower has already come under pressure from Washington to make use of this force before the weather closes in, lest it be seen as a hugely expensive white elephant. Some officers are even beginning to agitate for the huge numbers of transport aircraft sitting in England to be taken away and used as a stopgap supply line to support ongoing operations. For the airborne forces it might be a question of now or never.

As you reflect on Monty's new plan, Operation Market Garden, you wonder what to advise Eisenhower. You no longer have that much faith in Monty, but this does seem like a bold and dynamic plan and could at least win the Allies a Rhine crossing.

AIDE-MEMOIRE

* Operation Market Garden could deliver a great prize – a bridgehead over the Rhine.

* This is a very risky plan; inserting huge numbers of airborne troops sixty miles behind enemy lines has never been attempted before. Can it be done?

* The First Allied Airborne Army is a great asset waiting for its moment. Surely this is it?

* Backing Market Garden could also mean further delays to, or weakening of, the military offensives required to open up Antwerp.

* Without Antwerp's port operating as a major supply point, Allied operations in Western Europe will be seriously curtailed for many weeks or months to come.

* Can Market Garden be mounted quickly enough to make it viable, before the opportunity of such an audacious operation succeeding recedes?

* Intelligence reports imply that the Germans are pulling themselves together. To maintain any further momentum in the Allied advance you must act now.

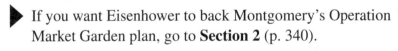

THE DECISION

Should Eisenhower back Monty's plan?

▶ If you want Eisenhower to back Montgomery's Operation Market Garden plan, go to **Section 2** (p. 340).

OR

▶ If you want to reject Market Garden, go to **Section 5** (p. 360).

SECTION 2
BRERETON'S CALL

1800, 10 September 1944, First Allied Airborne Army HQ, Sunninghill, near Ascot, England

A gathering of twenty-seven senior First Allied Airborne Army (FAAA) officers is about to begin a crucial meeting at the formation's headquarters in Sunninghill, Berkshire. The room is filled with anticipation and rumour. You are **General Lewis Brereton**, the recently appointed commander of the FAAA, though in truth this was not a post that you had relished. Until mid-August you were commander of US 9th Army Air Force in Normandy, an active and operational formation, whereas the position of commander of the FAAA has so far proved to be much more of an administrative one. You suspect that you have been eased out by General Omar Bradley, the leader of 12th US Army Group in France, but cannot prove it.* You do realise that you rub people up the wrong way at times, and you have been hauled over the coals by US Army Chief of Staff General George Marshall for your dalliances with the ladies whilst in the UK, but you are somewhat peeved by the position you now find yourself in.

In some ways you also believe that you are not being properly involved and consulted by the commanders on the continent. You have a team of airborne warfare experts, some with considerable experience, and yet operational commanders on the continent seem to plan

---

* It was true. Bradley commented, 'thank goodness' when Brereton was moved out of 9th USAAF. Another referred to Brereton as 'the fly in the ointment'.

---

without you. Nonetheless, this meeting promises to offer something tangible at last; so many airborne operations have been cancelled since you took up your post because the Allied armies in France have advanced at such speed, negating the need for your forces to be deployed. Now, however, progress has stalled and it could be the moment that your command is needed. As your staff assemble and the briefing begins, you realise you will have to make some key decisions.

2. Lewis Brereton

Lieutenant General Frederick Browning, the British deputy commander of the FAAA, begins the outline briefing. Earlier today he signalled from Belgium that a new full-scale operation, to include the FAAA, has been initiated at the behest of Field Marshal Bernard Montgomery, backed by Eisenhower himself. Browning had travelled over to the continent to confer with Lieutenant General Miles Dempsey, the commander of the Second British Army, which was advancing towards the Belgian-Dutch border. Monty had now instructed Dempsey and Browning to put together a plan that would crack apart the congealing front line and open up a route to the Rhine. Monty wants to use airborne troops in conjunction with Dempsey's Second Army to seize a path to the Lower Rhine at Arnhem, which at that moment lies some sixty miles behind enemy lines. The route to Arnhem passes through Eindhoven and Nijmegen and will require passage over a series of bridges spanning a number of rivers and canals.

The concept is not news to you; a similar plan has been bouncing around Sunninghill for a week already – codenamed Comet. But the scale of the new plan, Market Garden, is quite surprising and challenging. Comet had required the use of 1st British Airborne Division and the 1st Polish Parachute Brigade; Market Garden is altogether bigger and has added the deployment of two US airborne divisions – 82nd and 101st – to bolster the assault. As Browning continues to outline the plan that he and Dempsey have created, it becomes apparent that you have rather been cut out of the loop. Monty's great idea has been turned into an operational outline, without consulting you or your senior team. Browning and you have not seen eye-to-eye on a number of occasions already; only a few days earlier he had threatened to resign over a previous operational plan. You had called his bluff, he had backed down and the plan had fallen through anyway. Now it seems that Browning has seized the initiative and is obviously angling to take operational command of the new Market Garden plan.

You raise the issue of whether this will be a daylight or a night-time operation. After some discussion you decide on daytime, even though some concerns are raised about the vulnerability of the Allies' transport aircraft to enemy anti-aircraft gunfire, something that will be more of a problem in daylight. The tone of the meeting starts to shift, however, when Major General Paul Williams begins to talk. He leads IX Troop Carrier Command, the formation of transport aircraft and gliders that will deploy the airborne troops to the Netherlands. He points out that the airlift plan to get the airborne forces into the Netherlands, as imagined by Browning and Dempsey, simply will not work. The distances and times involved in flying to the Netherlands in mid-September mean that the increased number of troops for Market Garden cannot be deployed in one go. The distances involved, he argues, preclude the practice of each towing aircraft pulling two gliders, and the declining amount of daylight each day by mid-September means that only one flight to the Netherlands per day is possible. Therefore it will take three days to get everyone on the ground, even if the weather holds, which is hardly likely in north-west Europe in September.

Clearly the element of surprise, a fundamental element of airborne operations, will be lost if it takes so long to deploy. All the experience of airborne actions so far in the war has demonstrated that there is a vital window of opportunity in the first few hours of an operation for airborne troops to seize their objectives; after that, the enemy will react, create difficulties and slow everything down. Speed is therefore of the essence. But will there be sufficient force in the first wave to seize everything necessary to make Market Garden successful?

To add to the problems, many troops in the first wave will also have to be held close to the landing and drop zones (often some distance from the main objectives) to protect the arrival points for follow-up forces. This will reduce the numbers of troops that will be initially available to advance to the key objectives, mainly the bridges over the many rivers that lead the way to Arnhem.

The following day further meetings take place to discuss the air transportation plan. The use of precise, small-scale assaults – known as *coups de main* – directly onto the ground around the key targets, a similar tactic to that used on D-Day in Normandy at Pegasus Bridge, is ruled out (despite some opposition). Williams argues that due to issues with supporting escort air cover and the necessity for the Market Garden attack to take place in broad daylight, *coups de main* are too risky. Such small-scale assaults on the targets themselves will in any case immediately surrender the element of surprise and fully alert the enemy to the purpose of the assault.

Your Chief of Staff, Brigadier General Floyd Parks, updates you on progress and the growing worries from many staff. You grumble to Parks that if Monty had included you and your team in the initial planning of Market Garden (as Eisenhower had instructed him to do), all of these difficulties would have been made obvious and apparent. But a tipping point is now approaching. Williams and his staff are adamant that their modifications to the plan are absolutely necessary, but the changes they are insisting upon appear partially to compromise the airborne aspect of the whole operation.

Your options are difficult to accept. You could allow Market Garden to go ahead, despite your reservations: the gamble might still be worth it, considering the possible benefits; or you could go to Eisenhower and ask him to intervene. It would be quite incendiary to cut across Montgomery in this way; he has yet to forgive you for pulling the plug on a putative deployment of airborne troops onto Walcheren Island, which you and your team had regarded as being fraught with unacceptable risk. What should you now do?

AIDE-MEMOIRE

* Seizing a Rhine crossing will make the next stage of the war so much easier.

* The Market Garden plan is high-risk and breaks many of the lessons you have learned in the war up to this point.

* Market Garden could cause a collapse of Hitler's regime, if Allied troops start to pour into northern Germany.

* Surprise and speed of assault are fundamental to success in airborne operations – this plan sacrifices much of that.

* The FAAA is designed for this sort of operation – if it does not act now, when will it?

* Intelligence reports seem to indicate that the Germans are starting to recover. If you don't act quickly, the war could stagnate and drag on.

* If the plan fails, many thousands of Allied airborne soldiers could be isolated miles behind enemy lines.

THE DECISION

As you evaluate the evidence, what should you do?

▶ To back Operation Market Garden, go to **Section 3** (p. 346).

OR

▶ To go to Eisenhower and ask him to stop Montgomery's plan, go to **Section 6** (p. 366).

SECTION 3
ALL A QUESTION OF BRIDGES

15 September 1944, I British Airborne Corps HQ, Moor Park, north-west London

In your office on the second floor of I British Airborne HQ, you are deep in thought about Operation Market Garden, the forthcoming action that you helped to put together and are now about to lead. You are forty-seven-year-old *Lieutenant General Frederick 'Boy' Browning*, the dashing leader of I British Airborne Corps, deputy commander of the First Allied Airborne Army and a long-time pioneer and keen advocate of Britain's airborne forces. Married to the famous novelist Daphne du Maurier, you cut a smart, handsome and dapper figure; as one contemporary remarked, you are 'a lithe, immaculately turned-out man who gave the appearance of a restless hawk'. Never having seen operational command, however, you are keen to get into action before the war's end, but over the previous few weeks you and your colleagues have seen the continual cancellation of plans and operations; the ground war seems to be moving at such lightning speed that there is never enough time to prepare and carry out an air assault before the deployment is overtaken by events. You confide to your wife that you envy the commanders already in action, who have 'no worries except the battle in front of their faces'. Worryingly for someone like you, who harbours great hopes for reaching the top posts in the British Army, could the war end before you get into action? A lack of operational command could well hinder your progress.

Yet the latest scheme to deploy the airborne forces has thrown up a whole series of worrying problems for you. Just a few moments ago, in fact, you had a tetchy encounter with Major General Roy Urquhart, commander of 1st British Airborne Division, the unit of some 10,000 soldiers lined up to drop into Arnhem and seize the bridges across the Lower Rhine – the ultimate objective of Operation Market Garden. Urquhart announced to you, 'Sir, you've ordered me to plan this operation and I have done it, and now I wish to inform you that I think it is a suicide operation.' Then he walked out.

You are well aware of the growing concerns about Market Garden that had emerged since the first Sunninghill meeting. On 11 September at I Airborne Corps HQ you had met your senior operational commanders: Maxwell Taylor (101st US Airborne Division – 'The Screaming Eagles'), 'Jumpin' Jim' Gavin (82nd US Airborne Division – All American), Robert 'Roy' Urquhart (1st British Airborne Division) and Stanisław Sosabowski (1st Polish Parachute Brigade). With less than a week to pull the plan together, there was much to consider.

You had allocated the Eindhoven sector to the US 101st under the command of the highly respected and experienced General Maxwell Taylor. He quickly expressed his concern over the multiplicity of dispersed objectives that his forces had been asked to seize, which included bridges and the road south of Eindhoven over which British ground forces would advance in the opening hours of Market Garden. The thirty-seven-year-old Jim Gavin, the newly appointed divisional commander of the 82nd, also expressed his concern about the range of objectives he had been allocated. In addition to capturing the major bridges in Nijmegen itself, his forces would have to seize the very long and key bridge at Grave, at least one of four other bridges over the Maas–Waal Canal and capture the Groesbeek Heights, a pocket of high ground to the east of Nijmegen that dominates the approaches into the Netherlands from the Reichswald forest region of Germany. Gavin has to accept that his forces will be thinly spread and that 'a complete loss of control of the division might take place'. But what therefore are his priorities? The 82nd's tasks are further complicated because you have taken fourteen gliders of the first wave to fly in your corps HQ to the Nijmegen sector to coordinate the whole operation from the front line.

Yet Gavin's challenges are surpassed by those confronting Roy Urquhart, commander of the 1st British Airborne Division. He has been allocated 'the prize': Arnhem. But the tactical problems are even more pronounced. Urquhart's troops have to capture the available bridges in Arnhem, most obviously the large road bridge, hold landing and drop zones to enable follow-up troops to arrive – forces that could not be brought in during the first airlift – and secure the town itself, to allow Allied forces to break out beyond Arnhem once they had arrived across the river from the south. Sosabowski's Polish troops are to support the 1st Airborne and in particular secure the south bank of the river opposite Arnhem. Unfortunately, they are not planned to arrive until the second and third days of the operation, when all surprise will have been lost and the Germans will be fully aware of the unfolding operation. Sosabowski, never one to hold back his views, is less than impressed, particularly about what he sees as the casual attitude to the enemy's likely reaction.

The four officers depart to examine their particular parts of the plan in detail. Taylor, with his experience of coordinating and leading airborne operations in Sicily, Italy and Normandy fresh in his mind, takes his concerns to Lieutenant General Lewis Brereton (commanding First Allied Airborne Army, and Browning's boss) and insists that the targets be scaled down to what he believes are manageable proportions. Brereton and you grudgingly acquiesce. To complicate matters further, on 13 September a new intelligence report from Eisenhower's HQ emerges, stating that 'One thousand tanks reported in forest of Reichswald in Holland on 8th September, presumably a pool for refitting Panzer divisions'. However fantastical the information might appear, the potential threat from the area forces you and Gavin to think carefully about the plan for the 82nd. If the Germans have any significant offensive strength in the Reichswald and you lose control of the Groesbeek Heights, the success of Market Garden will be placed in jeopardy. Gavin is forced to consider lowering the priority of capturing the road bridge over the huge River Waal in Nijmegen itself. It can still be assaulted quickly on the first day, but now only by a single battalion of around 800 men that will have to rush some three to four miles to get there; it will be a big gamble and will break the basic concept that to capture a bridge quickly and effectively – especially one that was almost certainly primed with explosive charges by the Germans – it should be assaulted with alacrity, and preferably from both ends simultaneously. The risk is enormous: if the Waal road bridge cannot be captured quickly or, worse, is destroyed, surely Market Garden will fail?

Urquhart and his staff, however, are facing real and deeply worrying

Map 3: Gavin's plan for 82nd Airborne at Nijmegen

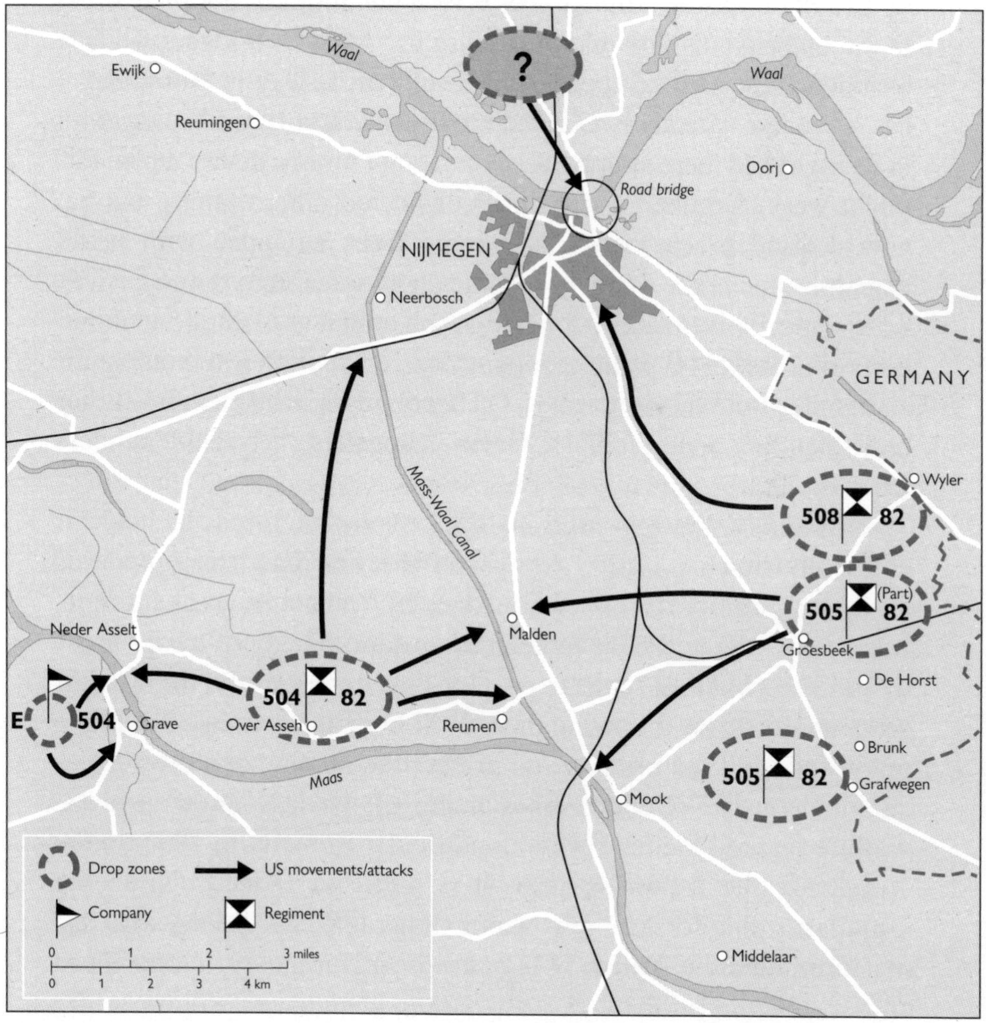

problems. The air forces and the air transport staff are insistent that any 1st Airborne drops and landings will have to take place some eight miles away from the bridge in Arnhem, a huge distance for the airborne troops to battle their way across; the Air Staff contend that anti-aircraft defences, urban terrain and unsuitable ground preclude dropping anywhere closer.

Only when the Germans' anti-aircraft guns between Arnhem and Nijmegen have been suppressed will the air forces acquiesce to fly low enough through the area to deploy paratroops south of the Lower Rhine, between Arnhem and Nijmegen. One option initially favoured by Urquhart – using a direct assault by a small force landing as close to the objective as possible to capture the Arnhem bridge quickly, a technique that yielded spectacular results on D-Day in Normandy – has, of course, already been ruled out because it will be too dangerous in daylight and there simply was not enough time to devise a plan that might work. Urquhart's team has considered other options and has now devised a scheme to land some jeeps equipped with heavy machine guns in the first wave (some eight miles from the objective), whose priority is to race to the bridge and hold it until three battalions, each of around 800 paratroopers, arrives on foot after traversing the eight miles from the drop zones. On hearing Urquhart's solution to the drop / landing-zone problem, Gavin is appalled: 'My God, he can't mean it,' he whispers to one of his team.

Ultimately, however, the inability of the air forces to land his entire division of 10,000 troops on the first day has scuppered Urquhart's ability to seize the bridges in Arnhem in great strength. Importantly, he will have to keep a substantial body of troops close to the drop / landing zones to provide cover and support for the subsequent airlifts on following days. Will Urquhart's troops be able to reach and hold the bridges and protect the landing and drop zones simultaneously? It contravenes many of the lessons of airborne warfare learned the hard way, in particular in Sicily in 1943 and in Normandy the following year. It is high-risk indeed. On seeing Urquhart's plan for Arnhem, Major General Richard Gale, who had masterminded the British D-Day drops in Normandy, is convinced

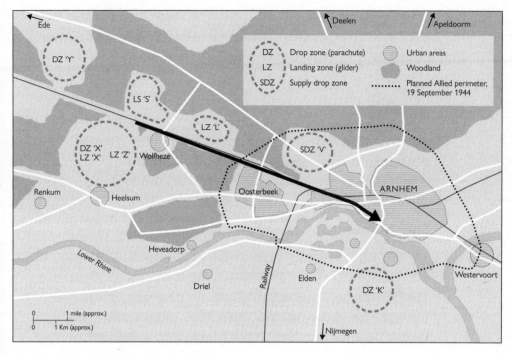

Map 4: Urquhart's Arnhem Plan

that it is so critically flawed that he would have threatened to resign rather than carry it out. You politely suggest that he keeps that to himself.

The manifest problems with Market Garden are profoundly worrying, as Urquhart's 'suicide' outburst demonstrates. All of your operational commanders have reservations about their own areas of the plan, but you can see the whole picture and you have seen how the original concept has been compromised. But by how much? The crucial difficulties facing Urquhart, and Gavin in particular, could be addressed by modifying the plan, choosing different drop zones and overriding the worries of the air forces. Urquhart could drop paratroopers south of Arnhem, despite the objections of the air forces; and Gavin could deploy some troops north of the Waal to seize the bridge straight away, although this could weaken control of the Groesbeek Heights. To push any of these changes through you might have to go to Monty or possibly Eisenhower himself to force Brereton and Williams to change their positions. And is there enough time to adapt the plan to meet the concerns of Gavin and Urquhart?

AIDE-MEMOIRE

* Gavin, Sosabowski and Urquhart are gravely worried about their parts of Market Garden. Is it prudent to go ahead when they have such concerns?

* The Air Staff and planners have made legitimate, intelligence-based arguments that cannot be ignored.

* The tactical necessity of deploying close to the bridges outweighs the concerns over terrain and enemy anti-aircraft gunfire.

* Dropping paratroopers south of Arnhem is a massive risk; heavy casualties could result.

* Drop zones miles from the bridges are simply too much of a gamble and risk disaster.

* Not capturing and securing the Groesbeek Heights threatens the whole plan. It has to be a priority.

* With just a few days to go, it is simply too late to change the plan.

* If you try to force through changes to the plan and the Air Staff refuse, the whole operation might be cancelled. Any chance of seizing a Rhine crossing would vanish and the possibly of ending the war quickly would be snuffed out.

THE DECISION

What should you do? Can, and should, you change
the plan now?

▶ To accept the risks as they are and proceed with Operation
Market Garden, go to **Section 4** (p. 354).

OR

▶ To insist on changes to the Operation Market Garden
plan, go to **Section 7** (p. 372).

SECTION 4
INTO BATTLE

1000, 25 September 1944, XXX Corps Tactical HQ, farmhouse north of Nijmegen, Netherlands

The news is not good and morale has slumped in the headquarters of XXX Corps, the formation coordinating the advance of the Allied ground forces towards Arnhem. For the last few days British and Polish troops had been battling the obdurate German forces between Nijmegen and Arnhem in order to try and reach, and relieve, the British airborne troops hanging on at the north bank of the Lower Rhine to the west of Arnhem. It has not been going well and now the tipping point has been reached.

3. Lieutenant General Brian Horrocks

You are the charismatic *Lieutenant General Brian Horrocks*, the commander of XXX Corps, a real 'Monty man' and a popular and effective battlefield commander. Having spent many months away from active duties due to severe wounds, you were called back into service by Monty in August to take over XXX Corps, which had rather lost its way. In truth you know you are not completely fit, and the stress of command in such trying circumstances is taking its toll. Nonetheless there is an important job to be done and you do not want to miss out. Just over a week ago, you issued a rallying cry to your troops as you prepared to start the ground assault of Operation Market Garden. Then you had likened the thrust towards Arnhem to a western movie, where XXX Corps (the cavalry) were there to rescue the 1st Airborne at Arnhem (the beleaguered homesteaders) from the bad guys (the Germans). Such theatrics now seem hollow and empty. The plan has unravelled, and now an overnight attempt to push troops across the Lower Rhine to the west of Arnhem, near Oosterbeek, has failed to achieve a great deal and has cost the lives of yet more Allied soldiers. The realisation is growing that Market Garden has failed.

Optimism for the operation did not last long, once it had begun on the beautifully sunny Sunday morning of 17 September 1944. More than 1,500 transport aircraft and close on 500 gliders lifted off into the skies over southern England and began the three-hour flight to the continent to begin Market Garden. Escorted and supported by 2,000 fighters and bombers, the vast air armada processed in relatively unmolested fashion to the drop zones in the Netherlands and, from just after 1230, gliders were soon bumping onto Dutch soil every nine seconds, whilst paratroopers began floating downwards from a mere 600 feet. The landings themselves were the most efficient and accurate of the war; in less than an hour, 19,000 troops were in position and largely on target, in contrast to many previous operations. Yet as successful as this had been, it still constituted only a little over half of the troops that had to be deployed; the others would be brought in over the next forty-eight hours. Would the initial landings be strong enough to win the day?

At 1430 on the same day British ground forces of your XXX Corps – composed of Sherman tanks, infantry in armoured transports, as well as infantrymen on foot, all covered by artillery barrages from 350 guns and RAF strikes from above – blasted their way out of the small bridgehead over the Bocholt–Herentals Canal on the Dutch-Belgian border and headed towards Eindhoven. It did not prove to be as straightforward as had been imagined. Soon the single road north to

Eindhoven was littered with wrecked vehicles and burning tanks. 'The Germans really began to paste us,' one commander reported. By the evening of the first day momentum was being lost. XXX Corps' assault quickly slowed down and failed to reach Eindhoven and link up with the 101st US Airborne Division on schedule. When they did arrive, they found that the Son Bridge had been blown up by the Germans, necessitating a delay while engineers built a replacement. When they got moving again, they made much more rapid progress and zoomed up the single road that was the only major route along which the advance had to progress, all 20,000 vehicles – tanks, half-tracks and trucks.

It was here that the serious flaws in the Market Garden plan began to be ruthlessly exposed. German opposition was much more intense than had been considered likely, and the intelligence reports proved to be rather wide of the mark; there was far more to contend with than simply old men and boys. The enemy's ability to react quickly and energetically to chaotic battlefield situations was well known and admired by Allied commanders, but had been rather passed over when Market Garden was put together. It seems that too much faith had been put in optimistic assessments that the enemy was 'weak, demoralised and likely to collapse' when attacked; this did not occur. And most of the initial opposition to the Allied landings was caused by scratch German forces, often poorly trained but determined, obdurate and well led by a core of experienced commanders. Jim Gavin's 82nd US Airborne Division performed very effectively, capturing the Grave Bridge intact and securing a route into Nijmegen. They also captured and held the Groesbeek Heights, although the threat from the Reichswald never materialised in anything like the strength anticipated; no signs of the 1,000 tanks, for example.

Yet in spite of everything that was secured, the initial limited push to the road bridge over the Waal in Nijmegen was rebuffed, as Gavin had feared, and the entire Market Garden plan hung in the balance. If the Germans had detonated the explosive charges placed on the bridge, it would almost certainly have sealed the fate of the British forces fighting in Arnhem, even if only some ten miles further on from Nijmegen. That the Germans held back from doing so opened up a door for the Allies to save the day.

Yet even this moment was squandered. The spearhead of XXX Corps arrived late into Nijmegen only to be informed by Gavin's troops that the Waal bridges remained in German hands. It took two more days of bitter street-to-street fighting by the American and British forces to batter their way to the river, but even then the bridges remained out of reach. This necessitated one of the most courageous and dynamic acts of the war, when 260 American troops in twenty-six open boats rowed frantically across the huge River Waal in broad daylight and assaulted the defended north bank. Around half of the US troops were killed or wounded but, in conjunction with intensive support from British artillery, they created the opportunity for British tanks to push north over the bridge.*

You all hoped that Arnhem might now be reached, but it proved not to be. The bridge at Nijmegen had only been secured as night was falling on 20 September, but you did not know what lay ahead of your leading forces, which were in any case too few in number to press on without proper support, particularly in darkness. Your tanks would have been sitting ducks, partly because much of the terrain between Nijmegen and Arnhem consisted of roads seated on embankments, which precluded manoeuvring by tanks and offered excellent defensive positions for the enemy. The American troops who had suffered so much forcing the river crossing were apoplectic: 'What in the hell are they doing? . . . All they seem to be doing is brewing tea!' one US commander raged. By the time XXX Corps got moving again the following day, the Germans had rushed units into the line ahead of you. Then everything got bogged down into a slow attritional slog, inching closer to the Polish airborne troops now arrived in Driel on the south bank of the Lower Rhine, with Arnhem beyond.

In truth it probably mattered little, for Urquhart's forces in and around Arnhem had long since lost control of the situation. His plan appears to have unravelled almost immediately from the moment the landings began. The fast jeep tactic collapsed; only one battalion of around 800 paratroopers (out of three that were supposed to get there)

---

* The second of which was commanded by Peter Carrington, later Foreign Secretary in Margaret Thatcher's government.

reached the bridge, where they hung on grimly until 21 September; radio links broke down; and Urquhart himself was cut off and out of communication with his HQ for some eighteen hours. German opposition was much more resolute and dynamic than had been imagined in the planning, and the British troops made very little headway, eventually being forced back into a defensive position around Oosterbeek, some four miles west from the bridge at Arnhem. The Poles, much delayed by poor weather, have been able to do little more than simply hold their position south of the river.

Yesterday you held a tetchy meeting with Sosabowski, commanding the Polish forces, and you clashed over if, and how, you could push some troops across the Rhine to aid the beleaguered 1st Airborne, now besieged at Oosterbeek. Now that plan has failed and you suspect that Market Garden needs to be terminated. You confer with Browning, commander of I Airborne Corps, who reminds you of a communication that he has received from Urquhart in Oosterbeek (see Source 1).

```
Must warn you that unless physical contact is
made with us early 25 Sept it is unlikely we
can hold out long enough. All ranks now
exhausted. Lack of rations, water, ammunition
and weapons. Even slightest enemy offensive
action may cause complete disintegration.
Have attempted our best and will do so as
long as possible.
```

Source 1: Urquhart to Browning, 24 September 1944

You then receive a call from your commanding officer, General Miles Dempsey, leading the Second British Army. You use code to discuss the options: 'Little One' meaning withdrawal of 1st British Airborne back across the Lower Rhine, and 'Big One' meaning further crossings. 'I think it must be Little One,' you say. 'I was going to tell you it would have to be the Little One anyway,' Dempsey replied. It is time to begin enacting Operation Berlin, the plan to evacuate what remains of 1st Airborne back across the Lower Rhine.

AFTERMATH

Ironically, Operation Berlin was one of the few aspects of Market Garden that went largely according to plan. Overnight on 25–26 September a little under 2,000 Allied troops escaped back to safety, but some 8,000 more had been lost, killed, wounded or captured. For the civilians left behind in Arnhem and Oosterbeek an appalling winter of hardship and privation followed. In order to fortify the north bank of the river the Germans forced the local population from their homes, with no regard for what happened to them. The wider population of the Netherlands also paid a high price for the failure of Market Garden; the Germans enforced a range of reprisals, including cutting off supplies of food and fuel, causing the infamous 'hunger winter'. It would take until the spring of 1945 for Allied ground forces to finally liberate Arnhem.

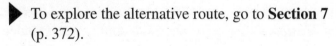

THE DECISION

▶ To explore the alternative route, go to **Section 7** (p. 372).

OR

▶ To know more about Operation Market Garden, go to the **Historical Note** on p. 377.

SECTION 5
THE ROAD TO WESEL

21 September 1944, Second British Army tactical HQ, Genk, Belgium

Second Army HQ is a hive of activity as officers bustle about preparing reports and briefings in preparation for the launch of Operation Naples III. Over the next few hours Allied forces will begin a major offensive from north-eastern Belgium towards the Meuse, and then on to the Rhine at Wesel in Germany itself. Lieutenant General Richard O'Connor's VIII Corps will spearhead the ground offensive breaking out towards Weert, whilst I British Airborne Corps begins landing in Venlo to seize crossings over the River Meuse. You are *Lieutenant General Miles Dempsey*, the quiet and unassuming commander of Second British Army. Considered, thoughtful and analytical, you avoid the limelight and attention – something you leave to Monty – and are regarded as a shrewd reader of terrain and as an intelligent planner. Operation Naples is a plan that you had backed and preferred over the abortive Market Garden. You had always believed that a push towards Arnhem was too risky and divergent a path away from the American armies in the south, with whom you need to cooperate closely. Your preference for an offensive to cross the Rhine at Wesel is still ambitious, but faces fewer river obstacles. With airborne support from Browning's I Airborne Corps and a well-prepared ground offensive, you have high hopes for your operation.

You got the go-ahead for Naples III in the wake of Eisenhower's decision to reject Operation Market Garden at the meeting with Montgomery in Brussels on 10 September. Monty had been furious, for his whole concept for driving Allied strategy in the direction that he saw as most likely to deliver immediate success had been blocked. Without the use of the First Allied Airborne Army, the option of reaching to the Rhine at Arnhem in September 1944 had gone; Ike had, in effect, enforced his policy of a wider push to the Rhineland in the hope of securing crippling losses on the enemy. 'Eisenhower himself does not know anything

4. Miles Dempsey (left) with Monty

about the business of fighting the Germans,' Monty fumed.*

When Monty had signalled to you that Market Garden was a 'no', you had been quite relieved; you had never wanted the push to Arnhem, since intelligence had begun to indicate the increasing level of resistance likely to be encountered in the area. It perplexed you and your Second Army intelligence team at the time that Monty's staff had not raised similar serious reservations, but it seems as though they had. You later found out that many of Monty's senior staff officers were worried about the choice of Arnhem. You then began working on alternative plans for the following phase of the campaign; you planned to convince Monty that a push to Wesel, and ultimately Cologne, to link up with Lieutenant General Courtney Hodges' First US Army and isolate the Ruhr was the priority.

Lieutenant General Frederick Browning, commanding I British Airborne Corps, was understandably dismayed by the cancellation of Market Garden, but when you pressed him for his thoughts on your Wesel operation he became more upbeat – he once again had great hopes of getting into battle before the war's end. Lewis Brereton, Browning's boss at First Allied Airborne Army, was also supportive: he too did not want to see his command grounded.

---

* From a telegram sent by Montgomery on 10 September 1944 to Vice Chief of the Imperial General Staff, Archie Nye.

Yet the greatest frustration was felt by Montgomery. His views on the failure of Allied campaign strategy since late August, since in truth he had surrendered control to Eisenhower, were being borne out. To Monty, there was simply no grip on the situation and he fumed that political considerations were driving Eisenhower's decision-making, and not military principles. He fulminated to this extent in his regular letter to his boss in Whitehall, Field Marshal Alan Brooke, the Chief of the Imperial General Staff, on the evening of 10 September: 'Just when a firm grip was needed there was no grip. In fact unless something can be done about it pretty quickly I do not see this war ending as soon as one had hoped.' In Brooke he found an ally, because Brooke had little regard for Eisenhower's abilities: 'Eisenhower as a general is hopeless! He knows little if anything about military matters.'* Brooke, however, was a more discerning political animal and understood the diplomatic realities of the Grand Alliance; it was inconceivable for the British to insist on a strategy that took the initiative away from the Americans, who were, after all, bankrolling the war and, by September 1944, providing the great majority of troops and resources fighting in the West. Monty refused to incorporate such political niceties into his military approach, but he had to accept the reality. He fired off an incendiary message to Eisenhower on 12 September in which he vented his deep frustration with the situation: 'I do not believe we have a good and sound organisation for command and control.' Eisenhower did not react well and for the first time called Monty's bluff: 'If you feel that my conceptions and directions are such as to endanger the success of operations it is our duty to refer the matter to [a] higher authority for any action they choose to take, however drastic.'**

Monty backed down; he knew there was little doubt that Brooke and Churchill would not have been willing to save him in such a situation; the agreed coalition position was to hand over strategic matters to Eisenhower, and the Americans were increasingly calling the shots now, so they could not easily be crossed. However much Montgomery

* A quote from Lord Alanbrooke's diary.
** Montgomery to Eisenhower, 10 October 1944, and his reply of three days later.

thought he was right – and he certainly did – it was a political battle he could not win. 'My dear Ike, you will hear no more on the subject of command from me,' he contritely acquiesced.*

Montgomery's climbdown appeared to have saved his job, but the chance of a full deployment of the First Allied Airborne Army had by now gone; both time and the practicalities were against it. Monty still believed that a push towards the Rhine was the priority, even if some were agitating that he should devote his energies towards opening up the port of Antwerp, which was still unusable as the Germans occupied the north bank of the Scheldt estuary. Until they were evicted, Antwerp was worthless.

When you delivered the Wesel-operation idea, with Browning's support, Monty saw an opportunity to deliver success, although it would mean close cooperation with the American armies too, something he had hoped to avoid by heading to Arnhem. Eisenhower and his staff expressed some reservations about resources being focused away from the Scheldt and opening the port of Antwerp, but agreeing to Dempsey's offensive seemed politically sound following the confrontation with Monty. Bradley had grudgingly acquiesced because he simultaneously secured the use of a considerable number of transport aircraft to strengthen the supply line reaching his forces in the south, and because the Wesel operation would result in closer cooperation with Hodges' First US Army.

Now the moment has come and, as the offensive begins and you see the paratrooper transport aircraft flying overhead, you are hopeful that the plan will deliver the level of success necessary to reach the Rhine and maintain British control of Allied strategy. Equally, you believe that the Wesel plan will bring closer cooperation with the American armies, as Hodges' forces will be flanking your main assault.

---

* A scenario that saved Monty's job later in the campaign, when he again overstepped the mark with Eisenhower.

AFTERMATH

Because of the lack of time for detailed planning, the airborne element of the operation was more cautious than the risky Market Garden and produced good results, and the initial ground offensive made some headway. The airborne troops captured their main objectives in Venlo, and hopes rose of a significant success. German opposition intensified, however, and eventually the offensive fell short of reaching the Rhine itself.

Nevertheless, Browning, Urquhart and Sosabowski all came out of the more limited airborne operation with their reputations intact. Montgomery was then forced to turn his attention to the problem of opening up Antwerp – a telegram from Ike in early October forcing his hand. But by then it was far too late to secure the full use of the port until November; indeed, the way to achieve this quickly had been lost many weeks previously. If there had been a moment to win the war in the autumn of 1944, it had long since passed and the war would drag on until the spring of 1945.

THE DECISION

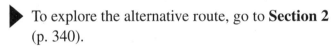

▶ To explore the alternative route, go to **Section 2** (p. 340).

OR

▶ To know more about Operation Market Garden, go to the **Historical Note** on p. 377.

SECTION 6
TURNING TO THE SCHELDT

1900, 15 September 1944, Montgomery's 21st Army Group tactical HQ, Belgium

Lieutenant General Frederick Browning, commander of I British Airborne Corps and a key architect of Operation Market Garden, which is due to begin in less than two days, has just arrived at Monty's HQ. He is the bearer of bad news and is clearly expecting a frosty and difficult meeting. Browning has had some frustrating discussions with planners at First Allied Airborne Army, then with his own intelligence team, and finally with his divisional commanders over the last few days and has become doubtful about the viability of the whole Market Garden plan. The latest intelligence reports from Eisenhower's staff and from internal sources at I Airborne HQ (including some new airphoto reconnaissance images) have also cast a shadow over the plan. Browning has steeled himself to tell Montgomery.

You are **Brigadier Edgar 'Bill' Williams**, 21st Army Group's thirty-one-year-old chief of intelligence, and you happen to be at Monty's tactical HQ when Browning arrives. Although you are not a career soldier (you are in fact an Oxford don), you have been one of Monty's key advisors since 1942. Generally he trusts your opinion on intelligence matters, but when you raised some concerns about the Arnhem plan, you – like many other senior staff at 21st Army Group – had been rather brushed aside. Now you can see that Browning too is worried about Market Garden.

Monty asks that you remain for the meeting. Browning outlines his objections to Market Garden, citing the new intelligence that the

Allies have received and the significant operational and tactical weaknesses now apparent in the plan; he shifts the blame onto Brereton. Montgomery is furious, but when you confirm that 21st Army Group intelligence matches that of the FAAA, he starts to accept the situation. You are mightily relieved, as you know that all the senior staff at 21st Army Group HQ are gravely concerned about

5. Edgar 'Bill' Williams

Market Garden. Monty eventually concedes that, in the face of opposition from his airborne commanders, the growing lack of enthusiasm of his own 21st Army Group staff and worrying noises emanating from Ike's command, Market Garden cannot now go ahead. He fires off a message to Lewis Brereton, commander of the FAAA, requesting that he adapt the plan along the lines that Browning has outlined, but to no avail. Monty disparagingly refers to Brereton as 'another gutless bugger; I have no use for him.'*

Operation Market Garden was formally cancelled the following morning, and many breathed a sigh of relief.

AFTERMATH

Monty fumed and tried once more to persuade Eisenhower to release control of the First US Army to him to bolster a new northern drive. Eisenhower demurred and pressed home the necessity of opening up Antwerp as soon as possible to ease the logistical paralysis that was consuming the Allied forces in Western Europe. If the push to the Rhine was now shelved, Ike argued, then Monty's priority had to be

---

* A criticism that he in fact levelled at Air Vice Marshal Trafford Leigh-Mallory in June 1944 when he blocked a previous airborne operation near Caen, Normandy.

Antwerp. In effect he forced Monty's hand with an unusually clear and almost pointed directive on 22 September; all efforts of 21st Army Group would be focused on ensuring the use of Antwerp.*

Montgomery wrote bitterly to Field Marshal Brooke, his boss in Whitehall, that the opportunity to win the war quickly had been thrown away. Of Eisenhower's leadership he wrote: 'He is completely out of touch with what is going on; he tries to win the war by issuing long telegraphic directives. The war may now go on into the winter.' Brooke and Churchill could not easily intervene; they had appointed Eisenhower, for good or bad.

The cancellation of Market Garden resulted in the redeployment of much of Dempsey's Second British Army away from the Belgian-Dutch border and over to the River Scheldt, and specifically the river estuary, much of which was still occupied by elements of the German Fifteenth Army.

In order to open up Antwerp, both banks of the river will need to be cleared of the enemy. Admiral Bertram Ramsay has been pressing for this to be prioritised since Antwerp had been seized intact on 4 September, but while Monty's eyes had been on Arnhem, German forces had been falling back into the Netherlands, retreating northwards across

Map 5: The Scheldt Campaign, 27 September 1944 – 7 October 1944

---

* A measure that Ike did not in fact take until early October.

---

the Scheldt to escape encirclement. Nonetheless, by the time of Market Garden's cancellation, they still retain significant forces on both sides of the Scheldt. Ramsay is nevertheless pleased that Monty's attention has at last been diverted to the problem.

A new major two-army plan, codenamed Operation Infatuate, is devised, employing naval and amphibious support. The First Canadian Army commanded by Harry Crerar will focus on clearing up the Breskens Pocket (south of the Scheldt estuary) while the Second British Army, under Miles Dempsey, will strike north and west from the Antwerp sector with the aim of capturing the peninsula Beveland. An ambitious second operation would follow to capture Walcheren Island, on which Browning's I British Airborne Corps will be deployed in conjunction with a series of amphibious commando forces to overwhelm German resistance. An idea to bomb the island of Walcheren and open its flood defences to hinder the defenders is rejected, and concerns over the deployment of airborne troops in a coastal region are headed off by Browning, who is now determined to seize this opportunity to get into battle, despite the severe objections of Brereton and some of his staff at the FAAA. Initially Brereton had written: 'I refused Operation INFATUATE because of intense flak on Walcheren, difficult terrain which would prevent glider landings, excessive losses likely because of drowning . . . and the fact that the operation is an improper employment of airborne troops.' Eisenhower eventually stepped in to support Monty, Browning and Admiral Ramsay in forcing the operation through.

German strength had grown in the area since early September, but the advantage lay very much with the Allies. Operation Infatuate began on 27 September and initial progress was significant; within six days the Canadian Army (including Polish and British forces) had cleared the Germans south of the Scheldt estuary. Dempsey's offensive initially encountered greater resistance, but soon pushed into Beveland and the assault on Walcheren Island began on 3 October. Browning's airborne forces suffered not-inconsiderable losses, with a great deal of confused fighting. Many paratroopers and glider forces came to grief in open waters, as Brereton and his staff had predicted, but the combined air, land and naval operation soon overwhelmed German defenders, who surrendered on 7 October.

Mine-clearing operations began shortly afterwards and Antwerp was declared open on 21 October. The supply situation in north-west Europe was duly eased, but the war could not be won in the autumn of 1944, much to Monty's chagrin. He continued to rail against Eisenhower's strategic leadership, but was ultimately forced to toe the line. Allied military operations in the autumn of 1944 were bolstered by the improved supply situation from Antwerp; and the later German counter-offensive in December, sometimes referred to as the Battle of the Bulge, was easily rebuffed. The Allies went on the offensive again in early January 1945, crossed the Rhine on 2 March, and Monty took the surrender of the Germans on 13 April 1945. Browning, Urquhart and Sosabowski, with the broadly successful Walcheren Island airborne operation under their belts, took part in the Rhine Crossing in 1945. Yet although Browning's military career continued into the 1950s, he never became Chief of the Imperial General Staff, which many believed he had always desired.

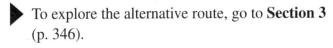

THE DECISION

▶ To explore the alternative route, go to **Section 3** (p. 346).

OR

▶ To know more about Operation Market Garden, go to the **Historical Note** on p. 377.

SECTION 7
THE BATTLE FOR THE BRIDGES

15 September 1944, First Allied Airborne Army HQ, Sunninghill, near Ascot

General Lewis Brereton, commanding officer of the FAAA, is holding an impromptu meeting with two of his key staff: General Paul Williams, commander of IX Troop Carrier Command, and you, ***Brigadier General Floyd Parks***, Brereton's forty-eight-year-old Chief of

6: Brigadier Floyd Parks

Staff. Despite entering the US Army in 1911 as a private, you had made steady progress towards higher posts, often in staff rather than command positions, and a few weeks ago you had taken up your current post at the FAAA. Having witnessed the ongoing frustration over all the recent cancelled operations, you are aware that many of your colleagues and operational commanders want to get into battle before the war passes them by. But at what cost, you wonder?

Now comes news that the latest scheme, Operation Market Garden, is also in peril. Brereton tells you and Williams that Frederick Browning, deputy commander of the FAAA and the man lined up to lead the airborne forces in Market Garden, has thrown down the gauntlet. Over the last few days the original plan has been clipped, changed, altered and modified to such an extent that Browning, reflecting the views of his commanders (including Max Taylor and Jim Gavin, two very highly respected American airborne divisional commanders), is threatening to pull the plug on the whole operation. He claims that unless Gavin and Urquhart are allowed to deploy paratroopers much closer to their objectives in order to seize the key bridges over the River Waal and the Lower Rhine immediately upon landing, the plan will have to be cancelled. Browning even said he would resign rather than carry out the operation as currently planned, although you know that, based on recent experience, this is probably a hollow threat. Montgomery has also been lobbying for changes to be made to the plan to accommodate Browning's concerns.

You, Brereton and Williams are stunned. Any attempt to fly through the area between Arnhem and Nijmegen to deploy troops in the area that Browning is now insisting upon risks heavy losses of aircraft to German anti-aircraft gunfire. In addition, the terrain to the south of the Arnhem bridges is also considered dangerous for airborne drops. Williams is furious – this is all the result of the inexperienced Browning cooking up the Market Garden plan with Miles Dempsey (commander of Second British Army), who has little understanding of airborne operations. If they had included FAAA staff in the original planning process, as Eisenhower had asked Montgomery to, none of this would have happened, Williams growls. Brereton agrees, but points out that if you do not go through with Market Garden, the FAAA risks being grounded for the winter; or, worse, the troops being deployed to the continent as regular frontline units and the transport aircraft being reassigned to supply duties.

Despite Williams' protests, Brereton tells you to explore how to accommodate the changes Browning wants while he confers with Eisenhower. At breakneck speed the alterations are implemented. Lieutenant Colonel John Frost's 2nd Battalion of the Parachute Regiment is selected to drop south of Arnhem, whilst the US 3rd Battalion of the 508th Parachute Infantry Regiment, commanded by Colonel Louis Mendez, will land north of Nijmegen. Additional Allied

Map 6: Modified plan for Arnhem and Nijmegen

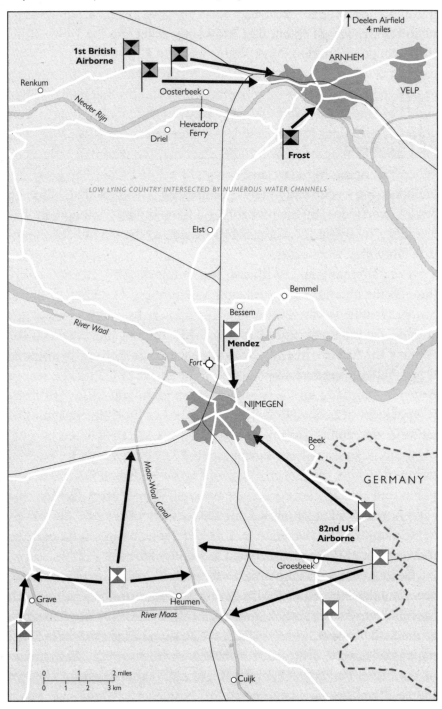

bombing missions are switched to support the new plan by attempting to suppress German anti-aircraft guns in the area. Allied air force planners are furious about the changes, but when Ike tells Brereton to make Market Garden happen, they are left with little choice but to carry on.

AFTERMATH

When Operation Market Garden began on 17 September the revised elements of the plan yielded some benefits at only limited cost, for the RAF's worries over the level of flak proved to be largely unfounded.* Mendez's troops rapidly descended onto the Waal road bridge, which was defended by just eighteen soldiers, and although the bridge was wired for detonation, the Germans failed to react quickly and the moment was lost. Mendez then waited for the rest of his regiment to arrive from the south of the river, but they were slow to advance and were soon blocked. Resistance hardened in Nijmegen, but once Gavin's forces had secured their other key objectives and it became clear that the level of opposition from the Reichswald in no way matched the pessimistic intelligence reports, the 82nd Airborne Division turned their attention to linking up with Mendez's now hard-pressed forces, still obdurately hanging on defending the road bridge in Nijmegen. German forces fought desperately, however, and Mendez's forces were almost driven off the bridge on two occasions before the rest of the division's troops cut through to link up in the early hours of 19 September.

Although XXX Corps was delayed in reaching Nijmegen, they did so later the same morning and crossed the Waal on the road to Arnhem. Mendez's troops had managed to hold a small perimeter around the north end of the bridge, but German resistance blocking the route north was determined. Furious fighting ensued to try to break through to Arnhem to relieve the now-beleaguered British 1st Airborne Division.

---

* In reality this was the case in the actual operation.

The revised element of Urquhart's plan had initially gone well enough. Frost's troops had deployed effectively, with few losses to flak and the supposedly unsuitable terrain, and they advanced rapidly to seize the road bridge from the south. His troops swept across the bridge, surprising the German defenders, and by the early evening of 17 September he was in position awaiting the rest of his brigade. The rest of the division's experience was not so positive; a unit of jeeps, which had been intended to rush the bridge soon after landing, had been ambushed and the two parachute battalions approaching Arnhem from the west had been halted by scratch German troops who had reacted quickly, mustering to meet the threat. German strength rapidly increased and the British were soon pinned down, and eventually began to be driven back towards Oosterbeek. Urquhart's division was soon fighting for its life. In Arnhem, Frost's force was soon hemmed in on both sides of the bridge and was subjected to heavy German counter-attack. Frost had expected the Poles to be dropped south of the river to link up with the British on 19 September, but poor weather caused this to be delayed; and Frost's force, still trying to defend both ends of the bridge, was on the point of being overwhelmed when, late on the 19th, British ground troops, advancing out of Nijmegen, appeared over the horizon from the south. By the evening they had linked up with Frost's troops, but by then German forces had squeezed the northern end of the bridge to such an extent that attempted breakouts were impossible. The bridge and the foothold on the northern bank in Arnhem itself became a killing zone, with severe fighting and heavy casualties. The perilous position of the rest of the 1st Airborne, now pinned down in and around Oosterbeek, also required immediate support and British ground forces spent the next two days battling to reach the river to throw across a lifeline to Urquhart's imperilled troops.

The position at Arnhem improved little over subsequent days and although 1st Airborne's position was secured, no significant progress was possible. The route to Arnhem, stretching back some fifty miles to Eindhoven, was still constantly under threat; German opposition was much stiffer than had ever been imagined possible; and Monty's great hope of pushing out from Arnhem into Germany itself never materialised. The war would now clearly drag on into 1945.

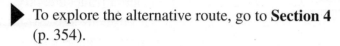

THE DECISION

▶ To explore the alternative route, go to **Section 4** (p. 354).

OR

▶ To know more about Operation Market Garden, see the **Historical Note** below.

HISTORICAL NOTE

Operation Market Garden is one of the most contentious and still-debated actions of the Second World War, partly because of the famous 1970s movie *A Bridge Too Far*, loosely based on Cornelius Ryan's book of the same name. The root causes of the failure of this audacious plan have been much discussed and have often centred on apportioning blame to commanders and planners and some of the particular mistakes they made. Perhaps the wider issue is that Market Garden really depended on everything working seamlessly, and of course in large-scale military operations that never really occurs. Its chances of success when it began were, therefore, very low indeed.

The historical path through this chapter is for Eisenhower to back the plan at his meeting with Montgomery on 10 September 1944 (from Section 1, go to Section 2). Ike was enthusiastic about Market Garden, although he just saw it as a means of securing a bridgehead across the Rhine; Monty hoped for much more than that, of course. Tedder was always likely to back it as it had a significant air element, was dynamic and innovative and opened up the possibility of injecting some momentum back into the Allied advance. Any reservations would have been based on his antipathy towards Monty or the fear that it would draw resources away from opening up Antwerp.

Market Garden began to unravel because of the initial disconnect in the planning process, and once the senior experienced airborne staff at the First Allied Airborne Army got their teeth into it, the weaknesses started to become apparent. Paul Williams, in carrying out his job, effectively did for the plan because he and his team pointed out that, in their opinion, the whole airborne force would take two to three days to deploy, and that made the whole scheme dubious in the extreme. Without *coups de main* assaults to seize the key bridges quickly, and with the air forces refusing to fly between Arnhem and Nijmegen because of fears of anti-aircraft fire, the prospects of the plan succeeding receded even further. Historically, no one at this point, however, was prepared to block the plan and it carried on (from Section 2, go to Section 3).

Would Browning have pulled Market Garden? He had done just that with an earlier airborne plan, even threatening to resign if forced to go ahead. With Market Garden everything was stacking up against it and there were moments when Browning could have threatened to cancel the operation unless key changes were made. It is doubtful that he could have forced them through – historically, Monty tried to badger Brereton into dropping his objections, but he refused. And by dropping Market Garden, Browning might have damaged his long-term career prospects. Consequently the plan went ahead, driven by optimism and a desire to get the war over with quickly (from Section 3, go to Section 4). Additionally, the Germans had recently started firing V-2 rockets at England from the Netherlands, so this was used to justify the attack towards Arnhem, which, if successful, might well have brought this campaign to an end.

If, on 10 September, Eisenhower had blocked Market Garden, it would undoubtedly have provoked a serious confrontation with Montgomery, which, if Ike had taken it up to the Combined Chiefs of Staff, would have almost certainly resulted in Monty's dismissal. Later in the campaign Monty was on the brink of being cut down in this way, before others with greater awareness salvaged matters for him. One option, however, if Market Garden had been cancelled was Dempsey's preferred plan for an offensive aimed at reaching the Rhine at Wesel – Naples III (from Section 1, go to Section 5). Such a plan was still ambitious and, in all likelihood, in view of the Germans' recovery, unlikely to achieve major success, but it was not going to founder in the way Market Garden did and would have worked more effectively in bringing the British and American campaigns closer together.

Another option, if Market Garden had not gone ahead, was an

earlier and concerted campaign to open up Antwerp. The even better option would have been for the Allies to have focused on this right at the beginning of September when they first captured Antwerp, but Monty was determined – not without merit – on keeping the momentum of the offensive going. Opening up Antwerp by clearing the Scheldt (from Section 2, go to Section 6) would help secure and bolster the Allied position in the medium-to-long term, but Montgomery was hoping for a knockout blow against Germany, and a Scheldt campaign was not that. If Market Garden had, however, later been cancelled and Monty had been nudged into clearing the Scheldt estuary, it would not have been as straightforward as it might have been three weeks earlier. Historically, Antwerp did not come into use until the end of November 1944, and Market Garden taking up resources was a factor in this delay. An earlier focus on this objective, in late September, might therefore have secured an earlier conclusion to the war, though not by a great deal.

Finally, what would the consequences have been of deploying some of the paratroopers closer to the bridges at Arnhem and Nijmegen (from Section 3, go to Section 7)? Leaving aside the likely apoplectic response of the air forces if they had been forced to fly their vulnerable transport aircraft through what their intelligence reports indicated was a very dangerous area, would it have made that much difference? Probably not, for Market Garden was so flawed that even if Frost had secured the road bridge at Arnhem, and Mendez the main bridge over the Waal, allowing XXX Corps to reach the south bank of the Lower Rhine a day or two earlier, the prospects of a major breakthrough were still very low indeed.

Arthur Tedder eventually became Marshal of the Royal Air Force and Chief of the Air Staff, playing a key role in the Berlin Airlift. He retired from the RAF in 1951, but continued to fulfil a variety of high-profile roles in society. He never quite overcame his antipathy towards Monty, but his main criticisms were largely professionally rather than personally based.

Lewis Brereton rather evaded any serious responsibility for Market Garden and continued to play a role leading the Allied airborne forces for the rest of the war in Europe. He returned to the USA later in 1945, performing a variety of roles until retirement in 1948. His reputation was not helped by his self-serving memoirs.

Frederick 'Boy' Browning's military career never really recovered from the failure of Market Garden and although he continued to serve in various important roles until he retired from the army, he never became Chief of the Imperial General Staff, as he desired. His reputation was cruelly skewered by the film *A Bridge Too Far*, which unfairly heaped the blame for Market Garden solely on his shoulders.

Conversely, the same film portrayed **Brian Horrocks** very positively. 'Jorrocks', as he was known, managed to skirt around any damaging responsibility for military blunders in his career and later developed a media profile in the 1960s, appearing in television documentaries. He also served as Black Rod, a largely administrative role in the House of Lords, from 1949 until 1963. To alleviate boredom during debates he used to fill in his football-pools coupons.

Miles Dempsey, who was often in the shadow of Monty, nonetheless served in a series of senior command positions in the Second World War. Although some believed Montgomery occasionally bypassed him in the chain of command, Dempsey was in reality a highly regarded figure, responsible for a series of operations, including devising the ground plan for Market Garden. He retired from the army in 1947, never wrote his memoirs and instructed that his papers to be burned on his death in 1969.

Edgar 'Bill' Williams returned to his academic career as a historian at Oxford University after the war, co-editing the *Dictionary of National Biography* from 1949 to 1980. He became warden of Rhodes House in 1952 and worked for the United Nations in a number of roles until his retirement.

Floyd Lavinius Parks became commander of the First US Airborne Army in 1945, and subsequently served in a series of other important roles in Europe, the USA and the Pacific. He later played a major role in public relations for the army in the 1950s and was a keen amateur golfer.

8

THE BOMB:
COPENHAGEN, MANHATTAN
AND HIROSHIMA,
1941–45

SECTION 1
THE MEETING IN COPENHAGEN

17 September 1941, Copenhagen

Werner Heisenberg and Niels Bohr, two of the world's most renowned and famous scientists, are setting out on an evening walk in Copenhagen, the capital of Denmark, a nation that has been suffering German occupation since April 1940. Although it seems to be nothing more than an incongruous stroll and chat, it is in fact a meeting at which the future outcome of the Second World War hangs in the balance – a meeting that might strongly influence the development, or otherwise, of an atomic bomb both in Germany and the West.

Aside from Albert Einstein, Bohr and Heisenberg are the leading theoretical physicists of the age. Bohr, a fifty-six-year-old Dane, has enjoyed a long and highly regarded career. With a doctorate at twenty-six and appointed professor at the University of Copenhagen in 1916, Bohr later inspired the establishment of the university's Institute for Theoretical Physics, which opened in 1921. The following year he was awarded the Nobel Prize in Physics for his work in the field of atomic structures and radiation, which firmly established him around the world as a leading light in his field. The high-flying forty-year-old German physicist Werner Heisenberg, initially inspired and cultivated by Bohr, emerged as a leading thinker in the field of physics in the interwar years. In 1927 Heisenberg published his famous paper on the 'Uncertainty Principle', a key development in understanding quantum mechanics, and by the 1930s he was Professor of Theoretical Physics at Leipzig and, like Bohr, a Nobel laureate, with a school of thought and a group of fellow physicists centred around his work, his name and his reputation.

You are **Margrethe Bohr (née Nørland)**, Niels' fifty-one-year-old wife and key confidante. Born into a family of physicists, architects and mathematicians, you married Niels in 1912 and have raised with him a much-loved family. But you have also been crucial to Bohr's professional rise to fame, too, often helping with the drafting of articles and acting as a vital sounding board as Niels has thrashed out his theories and ideas. He has openly acknowledged your importance, writing of one occasion, 'I went to the country with my wife and we wrote a very long paper.'

1. Margrethe with her husband, Niels Bohr

For many years Bohr and Heisenberg had been close friends, but since the rise of the Nazis and the outbreak of the war they have drifted apart. In April 1940 the German occupation of Denmark caused an even more damaging rupture in their friendship. Heisenberg is no Nazi, it seems, but he is a German nationalist, apparently willing to overlook Hitler's manifold grotesque failings. Bohr, partly of Jewish heritage, simply cannot accept that and is appalled at the racist, ignorant and destructive attitudes of the Nazis.

Yet although Heisenberg is a leading light in German society and culture, his physics is viewed very suspiciously by the Nazis, to such a degree that his reputation was tarnished for a time. There has a been long tradition of anti-Semitism in Germany, a tradition that Hitler has tapped into, exploited and expanded enormously, but which nonetheless pre-dated Hitler's rise to power. The most prestigious scientific posts in Germany's universities had long been tacitly denied to Jews, who were historically restricted informally to areas of study considered of lesser value – one of these, ironically, being theoretical physics. The Nazis have dramatically accelerated and reinforced this approach since 1933, as they regard any science linked with people such as Einstein as 'Jewish science'. Goebbels has labelled Heisenberg a 'White Jew', and many scientists have fled the Third Reich to avoid persecution. Heisenberg, however, has remained, and by the outbreak of war his value to the Third Reich is outweighing his 'non-Aryan' science. He is now an important figure in Hitler's regime, particularly as he is running the theoretical-physics institute at the University of Leipzig.

For you and Niels Bohr, the early stages of the war proved isolating and, since the occupation, Niels and his colleagues at the Institute for Theoretical Physics in Copenhagen have come under close scrutiny, with their work and activities closely monitored by the Nazis. Niels has kept as low a profile as possible and has avoided any entanglements with the German occupiers, but he has been dismayed and appalled by the events and consequences of the unfolding war. Contact with colleagues elsewhere is greatly curtailed, resources are at a premium and of course, being half-Jewish, he has felt particularly vulnerable under Nazi jurisdiction.

You have had little contact with Heisenberg for some time, particularly since the war broke out, but a few days ago Niels received a letter from Heisenberg – he was about to visit Copenhagen and wanted to meet up with Niels. But what did he want? It had been with some trepidation, and after deep discussion with you, that Niels agreed to a meeting, though on quite specific terms. Heisenberg was in Copenhagen for a German-sponsored conference and arrived early, keen to touch base with his mentor and friend. When he learned that Bohr would not attend the conference, however, Heisenberg went to the Institute for Theoretical Physics to lunch with Bohr and his colleagues.

However, it did not go particularly well, it seems. According to Niels, Heisenberg was bumptious, full of Germany's victories, and unsurprisingly the disgruntled Danes did not take kindly to his referring to the occupation of Western Europe as merely 'unfortunate'. Matters have not been helped by the continuing success of Hitler's armies in the east – Leningrad is now besieged, German troops are advancing hard on Moscow, and the Soviet Union seems on the point of collapse. In Heisenberg's view, the impending defeat of the USSR will be no bad thing and, more importantly, as the Third Reich seems to be on the verge of victory everywhere, it serves everyone's interests to accept the reality and come to terms with it.

Niels relates all of this to you upon his return, but in addition informs you that Heisenberg wants to meet up this evening to discuss matters of great importance. You remind Niels that it is quite possible the Gestapo have

2. Heisenberg and Bohr

bugged your home, so perhaps it would be better if the two of them went out for a walk and chatted there, like they did on the old days?* Heisenberg duly arrives and the two of them head off, but they return quite quickly and Niels is distressed. He tells you that Heisenberg had begun to raise an issue of vital importance, a fundamental question about the future and atomic power, energy and weaponry: was it acceptable for a scientist to aid his country in time of war to achieve these ends, he had enquired of Bohr? Niels had been taken aback – was Heisenberg suggesting that he was working on atomic weaponry for Germany, he asked? Was such a weapon a realistic possibility? Heisenberg had apparently suggested it was, and had begun to press Bohr for his views, or even whether he had any knowledge of any Allied plans for such weapons. He had then produced a sketch diagram of a theoretical atomic device. Bohr had immediately returned home, with Heisenberg in his wake.

What should Niels do? Heisenberg will soon depart. Should Bohr discuss the matter with him to find out more? Perhaps he can talk him out of it? Or should he dismiss Heisenberg immediately? The ensuing discussion could shape the path to atomic power and even atomic weapons.

---

* After the war no one could agree where the meeting had taken place – maybe in Bohr's study, or on a walk around Copenhagen harbour, or in the woods outside Bohr's house.

AIDE-MEMOIRE

* Heisenberg might well be trying to draw Bohr into a discussion in order to pick his brains about theoretical issues.

* Bohr is not at all sure that an atomic bomb is feasible. Would it not be better to convince Heisenberg of this by discussing the science?

* How far down the track towards an atomic bomb has Heisenberg travelled?

* Heisenberg is a close friend and colleague, whom Bohr knows well: he might be a German patriot, but he is no admirer of Hitler or the Nazis' warped views on science.

* Is he trying to discover if the Allies have a plan to make a bomb? Bohr has been in secret contact with the Allies, but in truth knows little.*

* Maybe Heisenberg wants, or would be able, to slow the development of atomic weapons in Germany, particularly if the Allies are not developing such a programme at great speed.

* Heisenberg briefly mentioned building an atomic-energy device, a reactor, but Bohr thought this might still be a step towards an atomic bomb.

---

* Heisenberg apparently knew that Bohr had been in contact with the Allies.

* The world would be better without atomic weaponry. Surely Bohr would not want Hitler to gain such a weapon?

* Clearly Heisenberg is already taking a great risk by divulging to Bohr the very fact that Germany has been working on an atomic programme. Why would he do that unless he is hoping to open up a channel of communications to the West to control and limit atomic-weapons development?

THE DECISION

How should Bohr react? What should he do in response to Heisenberg's questions and line of conversation? He quickly asks for your view.

▶ If you want Niels to close down the conversation with Heisenberg and have nothing to do with his approach, go to **Section 2** (p. 388).

OR

▶ If you want Niels to find out more and discuss the matter with Heisenberg in greater detail, go to **Section 3** (p. 398).

SECTION 2
THE ROAD TO MANHATTAN

9 October 1941, the White House, Washington DC

A crucial meeting is about to get under way between President Franklin Roosevelt, Vice-President Henry Wallace, and you, *Vannevar Bush*, the dynamic and determined fifty-one-year-old head of the US Office of Scientific Research and Development (OSRD). It is a meeting to consider the progress of the secretive OSRD's S-1 Section, which is the Uranium Committee, a unit that has been studying the potential of atomic power or perhaps even weaponry. You are an engineer and inventor with a string of successes and developments to your name dating back to the First World War, but your real ability centres on acting as a highly effective science administrator or 'fixer'. You fully recognise that liaison between the scientific community and government is sometimes fraught, particularly in times of war, and have thus deployed your shrewd understanding of the processes to keep the interaction both harmonious and fruitful. When Western Europe began to fall under the control of Hitler's Germany in the summer of 1940, you met Roosevelt and, in a short fifteen-minute meeting, persuaded him to establish a unit (eventually the OSRD) under your leadership to bring the scientific community directly into technological and engineering projects relevant to national defence, such as radar, weapons and other cutting-edge technology. The development of atomic power and weapons has until recently been low-profile, but matters are developing apace.

Niels Bohr had alerted the West to his meeting with Heisenberg and the possibility that the Germans were intent on developing some form of atomic programme. But the British and latterly the Americans were already pushing such developments themselves, even though the USA was neutral. The British government and military had become interested in nuclear fission, the basis of atomic energy, back in 1939, but there had been a good deal of scepticism about the viability of an atomic bomb. Early calculations had indicated a vast amount of the rare element uranium would be required to power a weapon but, ironically enough, it was two German refugee scientists, Otto Frisch and Rudolf

3. Vannevar Bush

Peierls, then working at the University of Birmingham, who had changed all that. They had picked up on Niels Bohr's published theory that only the uranium-235 isotope (which made up less than 1 per cent of uranium, the rest being 238) caused fission, and they calculated that the critical mass of uranium-235 required to power a bomb was just one kilogram, not the vast amount of natural uranium previously thought necessary. They concluded that if uranium-235 could be separated from the much more prevalent uranium-238, then an effective bomb was possible. Frisch and Peierls secretly wrote up their conclusions in a memorandum in March 1940 (see Source 1 on p. 390) and this set in motion the race towards the atomic bomb.

The British identified the key issues that would have to be addressed in creating a bomb, but the scale of the task was daunting, and their resources were thinly stretched as it was. In August 1941 a thirty-nine-year-old Australian scientist, Mark Oliphant, had flown to the USA in an unheated bomber to explore what progress was being made there towards atomic weapons. Earlier in his career he had

The attached detailed report concerns the possibility of constructing a 'super-bomb' which utilises the energy stored in atomic nuclei as a source of energy. The energy liberated in the explosion of such a super-bomb is about the same as that produced by the explosion of 1,000 tons of dynamite. This energy is liberated in a small volume, in which it will, for an instant, produce a temperature comparable to that in the interior of the sun. The blast from such an explosion would destroy life in a wide area.

In order to produce such a bomb it is necessary to treat a substantial amount of uranium by a process which will separate from the uranium its light isotope (U235) of which it contains about 0.7 percent. Methods for the separation of such isotopes have recently been developed. They are slow and they have not until now been applied to uranium, whose chemical properties give rise to technical difficulties. But these difficulties are by no means insuperable.

It is quite conceivable that Germany is, in fact, developing this weapon. At the same time it is quite possible that nobody in Germany has yet realized that the separation of the uranium isotopes would make the construction of a super-bomb possible. Hence it is of extreme importance to keep this report secret since any rumour about the connection between uranium separation and a super-bomb may set a German scientist thinking along the right lines.

Source 1: The Frisch–Peierls Memorandum, March 1940 (edited extracts) worked with Rutherford, the man who split the atom, and Oliphant was now one of the key members of the British team examining atomic weaponry.

He was dismayed to find out that the Americans had simply filed British reports and research findings and done nothing with them. In despair, Oliphant met the Uranium Committee properly on 26 August 1941 and set them straight: it was all about the bomb, he announced, not nuclear power, and they would have to get fully on board now or it might be too late. Crucially Oliphant told them that the British did not have the resources necessary to produce the bomb, so the Americans would have to take it on, even though they were still neutral.

4. Mark Oliphant

The Americans were not ignorant about atomic power and weapons, however, merely less concerned. In early 1939 two Hungarian physicists living in the USA, Eugene Wigner and Leo Szilard – worried about the prospects of Germany developing atomic weaponry in the wake of a fission breakthrough achieved in Berlin – persuaded the world-famous scientist Albert Einstein, by then also resident in the USA, to sign a letter to President Roosevelt to urge an investigation into fission technology and research (see Source 2).

This letter had at least prompted the establishment of the Advisory Committee on Uranium, but it was lethargically led and made little progress. Wigner despaired that they 'often felt as though we were swimming in syrup', while Szilard believed that the lethargy between 1939 and early 1941 had caused a serious delay in the development of the bomb. When you, as head of the OSRD, took on ownership of the Uranium Committee in 1940 its work had seemed of little immediate value.

Albert Einstein
Old Grove Road Peconic,
Long Island
August 2nd, 1939

F.D. Roosevelt
President of the United States
White House
Washington, D.C.

Sir:

Some recent work by E. Fermi and L. Szilárd, which has been communicated to me in manuscript, leads me to expect that the element uranium may be turned into a new and important source of energy in the immediate future. Certain aspects of the situation which has arisen seem to call for watchfulness and if necessary, quick action on the part of the Administration. I believe therefore that it is my duty to bring to your attention the following facts and recommendations.

In the course of the last four months it has been made probable - through the work of Joliot in France as well as Fermi and Szilárd in America - that it may be possible to set up a nuclear chain reaction in a large mass of uranium, by which vast amounts of power and large quantities of new radium-like elements would be generated. Now it appears almost certain that this could be achieved in the immediate future.

This new phenomenon would also lead to the construction of bombs, and it is conceivable - though much less certain - that extremely powerful bombs of this type may thus be constructed. A single bomb of this type, carried by boat and exploded in a port, might very well destroy the whole port together with some of the surrounding territory. However, such bombs might very well prove too heavy for transportation by air.

The United States has only very poor ores of uranium in moderate quantities. There is some good ore in Canada and former Czechoslovakia, while the most important source of uranium is in the Belgian Congo.

In view of this situation you may think it desirable to have some permanent contact maintained between the Administration and the group of physicists working on chain reactions in America. One possible way of achieving this might be for you to entrust the task with a person who has your confidence and who could perhaps serve in an unofficial capacity. His task might comprise the following:

a) to approach Government Departments, keep them informed of the further development, and put forward recommendations for Government action, giving particular attention to the problem of securing a supply of uranium ore for the United States.

b) to speed up the experimental work, which is at present being carried on within the limits of the budgets of University laboratories, by providing funds, if such funds be required, through his contacts with private persons who are willing to make contributions for this cause, and perhaps also by obtaining co-operation of industrial laboratories which have necessary equipment.

I understand that Germany has actually stopped the sale of uranium from the Czechoslovakian mines which she has taken over. That she should have taken such early action might perhaps be understood on the ground that the son of the German Under-Secretary of State, von Weizsacker, is attached to the Kaiser-Wilhelm Institute in Berlin, where some of the American work on uranium is now being repeated.

Yours very truly,
Albert Einstein

Then in the spring of 1941 you were made aware of the details of the British programme and of their belief in the feasibility of atomic weaponry, alongside a letter that was forwarded to you showing that the Germans were also developing their own scheme.

PALMER PHYSCAL LABORATORY
PRINCETON UNIVERSITY
PRINCETON NEW JERSEY

14th April 1941

Dear Dr. Briggs
    It may interest you that a colleague of mine who arrived from Berlin via Lisbon a few days ago, brought the following message:
    A reliable colleague who is working at a technical research laboratory asked him to let us know that a large number of German physicists are working intensively on the problem of the Uranium bomb under the direction of Heisenberg, that Heisenberg himself tries to delay the work as much as possible, fearing the catastrophic results of a success but he cannot help fulfilling the orders given to him and if the problem can be solved, it will be solved probably in the near future. So he gave advice to us to hurry up if USA will not come too late.
    Yours very truly,
    R. Ladenburg

Source 3: The forwarded letter

Yet you are in a quandary. The Germans seem to be embarked on a fission programme and the British are convinced that an atomic bomb is eminently feasible, possibly within three years.* But the USA is still technically neutral and at peace, and Congress is suspicious of spending large sums of money on military research and development; and when it did, it should be on programmes likely 'to yield results within a matter of months, or at most, a year or two'. The science underpinning the atomic project is still tentative in the extreme or even merely theoretical: there has been little progress in enriching uranium; no clear indication that another theoretical synthetic element, plutonium, could be used effectively; and as yet no self-sustaining chain reaction to create energy or produce plutonium has been achieved.** The British might be convinced by the prospect of an atomic bomb, but seemingly they want the neutral Americans (under your direction) to fund a good part of it. Intelligence on the German programme is also very sketchy; in the midst of a brutal and widening war (the Germans invaded the USSR in June 1941) would the Third Reich be in any position to find the resources to build an atomic weapon?

Recently you commissioned a new committee to determine the likelihood of success of an atomic fission programme; its findings are mixed and hesitant. Energy is a possibility, but a deployable weapon much less so, the report has concluded. Your deputy, James Conant, who now has direct responsibility for the Uranium Committee, remains deeply sceptical and suspicious of the enthusiasm of some of the physicists. To him the whole scheme is all very speculative – 'Wasn't this a development for the *next* war not the present one?' he had mused.

You are therefore confronted by a dilemma. As you meet Roosevelt and Wallace, you could press for a major investment in atomic research, probably at the expense of other important programmes, or you could just keep things ticking over in order to see what develops. The USA is neutral, after all.

---

* Oliphant later claimed that one year was a possibility.
** Nor would it be until the end of 1942.

AIDE-MEMOIRE

* The British are still lobbying hard – in their desperate circumstances, why would they not – but are they to be trusted?

* Has the German programme really made that much progress?

* Resources for military research and development are tight in the neutral USA, and this project would absorb vast amounts of money and time that could usefully be pumped into schemes more likely to offer a safe return on investment.

* What are the Germans up to?

* Even James Conant, your right-hand man, remains doubtful about atomic weapons.

* Can the USA really risk *not* investing? The consequences of the Germans acquiring an atomic device before you, or even the British, do not bear thinking about.

* If you act prematurely by requesting backing for a programme that then stalls, you risk losing credibility and the confidence of Roosevelt.

* There is still lingering doubt about whether anyone should develop such destructive power at all.

THE DECISION

What should you do? Should you approach Roosevelt immediately and ask for major backing on atomic-weapons development or wait and see how the ongoing research progresses?

▶ To ask the President for a major expansion in the development of fission technology, go to **Section 4** (p. 404).

OR

▶ To hold off until the science is clearer and invest the available resources in radar and other new weapons, go to **Section 5** (p. 412).

SECTION 3
HEISENBERG'S WAR

Autumn 1957, Niels Bohr's study, Copenhagen

The world-renowned physicist Niels Bohr is agonising over the writing and possible despatch of a letter to his old friend the German scientist Werner Heisenberg. Bohr has recently been much dismayed by the contents of a newly published book by Robert Jungk entitled *Brighter than a Thousand Suns.* Jungk's book (first published in 1956 in German, and subsequently translated into Danish the following year) is the first account of the atomic-bomb project in the USA and Germany's own attempts in the Second World War to build a similar weapon. In it Jungk has quoted correspondence with Werner Heisenberg, in which Heisenberg recounted his thoughts on the famous meeting that he had with Bohr in September 1941. They do not accord with Bohr's memories at all, and this is what has riled Bohr no end. You are *Margrethe Bohr*, Niels' wife and close confidant, still helping him with his work and correspondence. You and your husband have been deeply upset by the statements made by Heisenberg

5. Werner Heisenberg

about the 1941 meeting; they in no way match your own memories, and you and Niels have been wrestling with the content of a letter that you are thinking of sending to Heisenberg to put the matter straight.

Heisenberg told Jungk that one of his main aims in wanting to meet Bohr in 1941 had been to cooperate over slowing down the pace of the atomic-bomb programmes in Germany and in the West. He claimed that Bohr had also been shocked and surprised that a bomb was even feasible. Jungk's book also argued that German scientists had deliberately slowed the progress of an atomic-weapons programme under Hitler. Upon reading Jungk's book, Bohr had been appalled. His recollection of the meeting with Heisenberg was markedly different and called into question the altruistic claims made (see Source 4).

Dear Heisenberg,

I have seen a book, *Brighter than a Thousand Suns* by Robert Jungk . . . and I think that I owe it to you to tell you that I am greatly amazed to see how much your memory has deceived you in your letter to the author of the book.

Personally, I remember every word of our conversations.

You and Weizsäcker expressed your definite conviction that Germany would win and that it was therefore quite foolish for us to maintain the hope of a different outcome of the war.

You spoke in a manner that could only give me the firm impression that, under your leadership, everything was being done in Germany to develop atomic weapons and that you said that there was no need to talk about details since you were completely familiar with them and had spent the past two years working more or less exclusively on such preparations.

Source 4: Bohr's letter to Heisenberg, 1957 (extracts)

Bohr went on to write that he had been shocked that Heisenberg was actually intent on developing an atomic bomb for Germany and that was why he had remained silent and had abandoned the discussion. He knew full well that a bomb was theoretically possible and had already published and lectured on the matter before 1941, but had no insight into progress in the West on such matters. The suspicion grew that Heisenberg was simply rewriting history to avoid contamination from the appalling legacy of the Third Reich. You and Niels have discussed the notion that Heisenberg might have tried to slow the German programme during the war and, after recalling his tone and comments in 1941, find it fanciful and disingenuous.

In truth, the German atomic-bomb programme – the *Uranverein* project – had not made that much progress, nor did it. In part this was down to a lack of expertise in the field of theoretical physics, a result of so many scientists in the field having suffered discrimination under the Nazis and having fled to other countries. As a result the *Uranverein* project, under Heisenberg's leadership, had failed to grasp some key fundamentals concerning how to make an atomic bomb. Heisenberg, for example, did not appreciate the amount of uranium-235 (U-235) required to form the critical mass necessary to start and sustain an atomic chain reaction. The German team was working on the assumption that the critical mass necessary was 1,000 kilos, far too large to make a practical bomb. U-235 could theoretically be isolated from U-238, but it would be technically very challenging indeed and monumentally difficult to mass the amount required.

U-235 was also necessary to start the process to create plutonium, another route to a possible atomic bomb. Because the German team had miscalculated the amount of U-235 to start a chain reaction, the plutonium method seemed to be the better route to nuclear power and, ultimately, a bomb. Yet the German team had erred in their thinking on this process, too. They had settled on the use of heavy water for the moderation process, necessary to control the reaction required for the creation of plutonium.* But heavy water was less effective for this

---

* Plutonium had been discovered in the USA in 1940, but the Germans also knew of the theory, though they called it a different name at the time.

than graphite, a material erroneously dismissed by Heisenberg's team. It therefore made the Germans reliant on heavy water from Norway, which could be targeted by the Allies.*

Heisenberg had always maintained that he knew about the diffusion calculation necessary to properly understand the amount of U-235 required to start a chain reaction, but chose not to do it, in order to slow Hitler's atomic-bomb programme. Yet secret recordings made by the Allies in August 1945 of Heisenberg in conversation with other captured German scientists implied otherwise – it seems as though he had simply failed to make the right connections and it was only when he heard that the first atomic bomb dropped on Hiroshima had been a U-235 device that he realised his error.

As you discuss the matter of the letter to Heisenberg with Niels, you reflect on what might have been at the meeting in 1941. What if Bohr had tried to dissuade Heisenberg about the atomic bomb and the matter of the diffusion calculation had come up? If Heisenberg had departed from Copenhagen, alerted unerringly by Bohr to the U-235 conundrum, could the Germans have made an atomic bomb?

If Heisenberg had focused on the matter, it would probably have been apparent that the amount of U-235 required was closer to 10–15 kilos, and not the vast and unfeasible amount he previously thought. Yet many other significant challenges would have had to have been overcome. The Nazi hierarchy remained deeply sceptical, and even if Heisenberg had made a better pitch for funding, it is still doubtful that the Nazis would have backed him. They were beguiled by Wernher von Braun and his rocket programme – science they could grasp more easily than the theoretical mumbo-jumbo of Heisenberg. Von Braun was a dynamic and consummate salesman for his scheme, in a way that Heisenberg could not match. Germany eventually spent more on the V-2 rocket programme than the USA did on the Manhattan Project, and von Braun was too shrewd a political operator to allow resources for his project to get channelled elsewhere.

Ultimately it is highly unlikely that, even if Bohr had unwittingly

---

* Historically a target of Operation Gunnerside in 1943, and depicted in the 1965 film *The Heroes of Telemark*.

flagged up the issue of the diffusion calculation, Germany would ever have been in a position to develop a bomb before the war's end. After all, even with the huge investment made by the USA, the atomic bomb was not ready in time to be used in the war in Europe.

Back in Copenhagen in 1957, you and Niels consider deeply the necessity of sending the letter to Heisenberg. Bridges have been built since 1945, and how much unpleasantness would be stirred up by openly contesting Heisenberg's account? The letter is drafted and redrafted, but is never sent.

AFTERMATH

The German nuclear-weapons programme all but folded in 1942, partly as a result of a conference held by Albert Speer in June of that year. Heisenberg, though indicating that a bomb might be feasible, established that it would take a monumental effort, with success only being possible many years in the future. Although it was contested for many years by Heisenberg, part of the issue was that he had not successfully calculated the amount of U-235 required to make a bomb, overestimating the critical mass necessary by a factor of one hundred. It made a uranium bomb all but impossible, although a plutonium bomb was still considered feasible, if only after many years of development and investment. The Third Reich chose to invest elsewhere and, after the purge of theoretical physicists by the Nazis, it may well never have had the expertise to make a bomb anyway, let alone one in time to affect the outcome of the war.

Heisenberg's reputation, salvaged by Jungk's book in 1957, slowly started to crumble. *Brighter than a Thousand Suns* was eventually discredited, even to the extent that Jungk himself disowned it in later life, claiming that the German scientists – Heisenberg in particular – had deceived him. Bohr's draft letters to Heisenberg, though never actually sent, survived and were eventually published in 2002, forty years after Niels Bohr's death.

THE DECISION

▶ To follow the alternative route, go to **Section 2** (p. 388).

OR

▶ To know more about the development and use of the atomic bomb, go to the **Historical Note** on p. 432.

SECTION 4
THE DECISION TO DROP THE BOMB

25 July 1945, Potsdam, Germany

A momentous meeting is taking place, one that could potentially change the nature of war for ever. President Harry S. Truman, in post for just a few months, is mulling over his options concerning the new, enormously powerful weapon developed by the USA, the atomic bomb. The responsibility for this weapon came his way when he stepped up to the highest office following the death of the great Franklin D. Roosevelt. Simultaneously, Truman is in the midst of the last

6. President Truman (left) with Henry Stimson

great multinational conference of the war, with both Stalin and Churchill in attendance. You are **Henry Stimson**, the seventy-eight-year-old
US Secretary of War, a tough, conservative Republican, suspicious of
Stalin and hostile towards communism. Roosevelt brought you into his
administration, despite his being a Democrat, to bolster the government's base of support, and because you are known as someone with
grip and drive, despite your advancing years.

The Allied conference is taking place at the Cecilienhof Palace in
Potsdam, a short distance from Berlin, Hitler's capital, which now lies
in ruins after its capture and sacking by Soviet forces. Truman, however, is meeting his senior team – James Byrnes (Secretary of State),
General George Marshall (Army Chief of Staff) and you – at his own
residence for the conference, a lavish house previously occupied by a
distinguished German publishing family, but which has recently been
seized by the Red Army. The matter at hand is the issue of the atomic
bomb and its deployment against Japan.

All the intense work on the atomic-bomb programme since the
autumn of 1941 has now paid off – the weapon is ready to use, having
just been successfully tested. Vannevar Bush, head of the Office of
Scientific Research and Development, secured Roosevelt's commitment to the atomic bomb at the crucial meeting on 9 October 1941,
and from then onwards presidential backing provided Bush with considerable kudos and status, which he deployed enthusiastically to cut
through red tape and bureaucratic torpor. Whenever it was required,
Bush had pointed to the White House's firm support and Roosevelt's
backing; it was usually enough for Bush to get his own way.

The atomic-bomb programme had subsequently been named the
Manhattan Project, and the army, rather than the navy, had been
selected to lead the project (there was no independent air force in the
USA until 1947). In the summer of 1942 you chose Brigadier-General
Leslie Groves to manage the programme, predominantly to inject grip
and drive into the operation. Groves had wanted a combat field command, but you would have none of it. He had been informed, 'The
Secretary of War has selected you for a very important assignment,
and the President has approved the selection. If you do the job right,
it will win the war.' Alongside his fierce determination, Groves' most

important contribution was to insist on the appointment of the Berkley scientist Robert Oppenheimer to lead the secret weapons laboratory – in effect the scientific team that would design the bomb. Concerns were expressed about Oppenheimer's very limited experience of large projects, and there were deep reservations about his 'left-wing politics'. Groves was very impressed by Oppenheimer, however, and was quite insistent on his appointment; as one colleague put it, 'it was a stroke of genius on the part of General Groves, who was not generally considered to be a genius'. You had backed Groves' judgement.

Groves' and Oppenheimer's grip and vision drove the Manhattan Project forward and, backed by Bush's funding and your oversight, tremendous progress was made. The method of dropping the bomb, however, began to emerge as a potential problem; these bombs were going to be substantial pieces of ordnance, and finding an aircraft with sufficient range or payload capacity was an issue. In 1943 existing American bombers simply did not have the capability to deliver an atomic bomb across the ranges necessary in the Pacific theatre. But then the new Boeing B-29 Superfortress seemed to offer a solution.

The B-29 did not see service until mid-1944, but was a state-of-the-art aeroplane that incorporated the most modern and cutting-edge design elements, making it the most advanced bomber aircraft then built. It was also the most expensive single military investment programme made by the USA during the Second World War, requiring some $3 billion, a figure far in excess even of that spent on the Manhattan Project. A new United States Army Air Force unit, 509th Composite Group, has been formed, alongside a conversion programme called 'Silverplate', in which some B-29s have been modified to accommodate atomic bombs.

By the spring of 1945 all the necessary work to prepare the atomic bombs for use was nearing completion. Yet the greatest issue still to be confronted was the decision to use the bomb at all. Planners had originally worked to the assumption that the bomb would be used on Germany, but the war in Europe had come to its final stages before the bomb was ready. This left Japan still fighting, so Roosevelt and his staff decided that all efforts should be directed towards deploying the bomb in the Pacific theatre, probably in the summer of 1945. No one wanted the war to drag on into 1946, and the bomb might be a way to prevent this.

Japan was already looking doomed to defeat, crippled by military and naval failures strangled by a blockade, and hammered from the air by the new B-29 bombers using conventional bombs. In March 1945, in one B-29 raid alone, it was estimated that some 100,000 Japanese people had perished in a horrific firestorm, worse than anything witnessed in Europe. Yet there were no clear and obvious indications from Tokyo that Japan was seriously thinking about surrender. Time was ticking down on the moment when the atomic bombs might be ready to use, and a list of possible targets had been drawn up, including Hiroshima, with Nagasaki included as a late substituted choice.

Matters had been complicated when Harry Truman became President on 12 April following Roosevelt's death. Truman had only entered high office as Vice-President in January, and had held just two private meetings with Roosevelt before FDR's demise; he was hardly 'in the loop'. Most importantly, Truman had not yet even been briefed about the existence of the Manhattan Project, which was still shrouded in the utmost secrecy. The sixty-year-old Truman had been a late choice to stand as Roosevelt's vice-presidential running mate in the autumn 1944 elections, mainly because he had been considered a 'safe' option or, to some people, 'the common man's common man'. Considered straight-talking, assertive and cocky, Truman was nonetheless dismissed by others as a 'small bore politician of country courthouse caliber only', despite his apparent homespun honesty.

You had briefed Truman about the Manhattan Project shortly after he became President, mentioning the atomic bomb as a weapon of 'almost unbelievable destructive power', but one that also brought foreign-policy benefits. You had pointed out that you believed there could be a strategic advantage to the USA in emphasising the atomic bomb as a diplomatic weapon against the USSR, potentially 'a master card'. Truman's newly appointed Secretary of State, James F. Byrnes, fervently supports this idea. Concern has been growing in Washington over Stalin's clear intention of imposing pro-Moscow governments in those Eastern European states recently liberated by the Red Army in 1944 and 1945. How could Washington get Stalin to accept self-determination for those countries? Maybe the atomic bomb, a weapon the Soviets did not yet have, could lever some concessions out of him? You have even suggested delaying the start of the Potsdam Conference to allow enough time to test the atomic bomb, so that the President might confidently use it as a diplomatic tool to intimidate Stalin.

Yet the previous day, when Truman told Stalin about the new weapon acquired by the USA, it seemed to have little effect; 'then use it' Stalin had stated with a shrug. The American team had even been unsure whether the inscrutable Stalin fully understood what he was being told.* Stalin did, however, confirm that the USSR would fulfil its commitment to declare war on Japan three months after the end of the war in Europe, meaning that a Soviet attack would begin on Japanese-occupied Manchuria on 8 August. 'Fini Japs when that comes about,' Truman whispered to you.

Defeating Japan was obviously the priority for Washington, and America's military chiefs had been conferring on how best to bring this about. Alongside continuing the naval blockade and the aerial bombardment of the Japanese home islands, the Joint Chiefs had begun preparing a plan for the invasion of the southern Japanese island of Kyushu, to be codenamed Operation Olympic, to start in November 1945. The military chiefs blanched at the thought of it, however; the scale of the operation would be huge and the likely casualties on both sides considerable. The recent battles in Iwo Jima and Okinawa had brought about heavy loss of life, and there is little reason to suspect that Japanese resistance will melt away if the US launches an invasion of Japan itself. As it is, the war is costing thousands of American lives every month; no one wants it to go on much longer, and no one relishes the thought of invading Japan.

But will the Japanese government in Tokyo see sense? Washington is insisting on unconditional surrender, as it has reaffirmed at Potsdam. But the Japanese government, even though now apparently willing to negotiate in July, appears far from accepting this. Your

---

* In fact Stalin was fully briefed, due to Soviet spying on the Manhattan Project.

intelligence staff have broken Japanese diplomatic codes and you know that the government and military in Tokyo are divided between peace and continuing the war, but there is no clear indication that anything approaching acceptable terms is likely to emerge. Yet Truman is desperate to end the war as soon as possible, to save American lives and to focus on global reconstruction. Might the atomic bomb be a factor in ending the war quickly? If so, how should it be used?

A group named the Interim Committee had been established by Truman in the spring of 1945 to evaluate the employment of the atomic bomb, and they have recommended that it should be used against Japan without warning, to maximise its effects and potentially shock Tokyo into immediate and unconditional surrender. Consideration has been given to a demonstration or a detailed warning about the bomb being issued to Tokyo, but this has been rejected (though not without dissent) as being likely to lessen the shock and undermine the bomb's effects. It is also true that the bomb might not work properly at all, and you only have just mustered the materials for two bombs, as it is.

As you assemble and begin discussing the matter, what should you advise the President?

AIDE-MEMOIRE

* The bomb could well bring the war to an
  end, thus saving many American lives.

* If you use the bomb, it seems certain that
  tens of thousands of Japanese civilians
  will die at a time when Japan appears
  willing to negotiate.

* The Japanese are seeking terms that are
  quite unrealistic, akin to a mutual
  ceasefire.

* Other options, such as the Soviet
  declaration of war on Japan, to initiate

the occupation of Manchuria on 8 August, could well push Japan to surrender without the bomb being used.

* The bomb could place a marker down to the Soviets - Stalin may not have grasped the theoretical implications of the bomb, but a demonstration would surely show American superiority.

* The Interim Committee has advised that use of the atomic bomb - alongside a Soviet declaration of war, the continuing blockade and conventional bombing - offers the best chance of forcing Japan's surrender quickly, and that is surely the priority.

* Is it necessary to stick to the unconditional-surrender demand? You did with Germany, but you negotiated with Italy.

* The combined impact of the bomb, the Soviet declaration of war and Japan's rapidly fading military capability might be needed to force the hardliners in Japan into surrender on acceptable terms. Any hesitation could prolong the war.

* While use of the bomb might intimidate the Soviets, it could also poison relations between Moscow and Washington and thus make the post-war world more difficult to manage.

THE DECISION

The pressure is growing on Truman. What do you advise him to do?

▶ If you want Truman to hold off using the atomic bomb in favour of forcing Japan to surrender by other means, go to **Section 6** (p. 422).

OR

▶ If you think Truman should give the order to use the atomic bomb against Japan, go to **Section 7** (p. 426).

SECTION 5
DOWNFALL

29 July 1945, US Joint Chiefs of Staff meeting, Washington

The most senior military body in the USA, the Joint Chiefs of Staff, is being confronted with sobering news that has hammered home some stark realities. A new intelligence briefing has just been submitted that highlights the alarming rate of Japanese military preparations and build-up on the southern island of Kyushu, the first target of the American invasion of Japan (codenamed Operation Olympic) being planned for the autumn of 1945. Japanese troop numbers on the island have increased dramatically from 80,000 to more than 200,000 and there is 'no end in sight'.

7. General George Marshall

This increased figure has rung alarm bells across Washington, and you, *General George Marshall*, the head of the US Army, are in despair, even to the point of contemplating the cancellation of Olympic. Prior to the new report, you and your fellow US military leaders had already been deeply concerned about the prospects of carrying out a direct invasion of Japan; you are dismayed by the human cost of the recent battles in the Pacific theatre. Close on 50,000 American soldiers have been killed in the previous twelve months in the fighting in the Philippines, the Mariana Islands, Iwo Jima and Okinawa, with four times that number suffering wounds, battle exhaustion and disease. If the Japanese prove able to mount serious opposition to Olympic, then American casualties could rapidly mount to desperate and worrying levels. And Olympic is only the first stage, with the larger Operation Coronet – a landing on the island of Honshu to capture Tokyo itself – pencilled in for the spring of 1946, both plans constituting Operation Downfall, the conquest of Japan.

Yet the other options for ending the war with Japan before Operation Olympic would be necessary are looking unlikely to prove fruitful. The ever-tightening blockade appears to be inflicting crippling damage

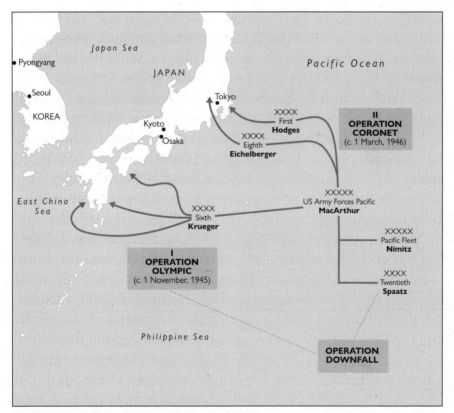

Map 1: Plan for Operation Downfall

on Japan, but there is no clarity or compelling evidence that it alone will force a surrender in the near future. The B-29 firebombing raids have killed in excess of 250,000 Japanese citizens since March, but still Japan's government remains obdurately, and perhaps irrationally, resolute. The leadership in Tokyo seems fixed on insisting on totally unrealistic surrender terms. In Washington you recently heard that a new weapons programme, based on some form of atomic power and which has been progressing since 1941, will not now be ready until the summer of 1946, if it works at all. The political leader of this so-called Manhattan Project, US Secretary of War, Henry Stimson, admitted that it had not been prioritised back in 1942, when resources had been channelled into other areas of research. Although the Man-hattan Project might ultimately yield a major military breakthrough, it can offer no immediate solution to ending the Asia-Pacific War in 1945.

All hopes are therefore pinned on the Soviet Union's impending declaration of war on Japan in early August. Truman's predecessor, Franklin Roosevelt, secured the agreement of Stalin for the Soviet Union to begin military operations against Japanese territories on mainland Asia, such as Manchuria (or Manchukuo, as Japan called it), three months after the end of the war in Europe – all in the hope of ratcheting up the pressure on Tokyo. The Japanese government had been trying to prod Moscow into acting as an intermediary with Wash-ington, so Truman, along with you and the other Chiefs of Staff, hope that the sudden rupture with the Soviet Union will be enough to shock Japan into an unconditional surrender, or at least something close to it.

When the Soviet attack came on 8 August, and with it the rapid collapse of the Kwantung Army defending Manchukuo, Tokyo was indeed thrown into confusion and despair. Yet it does not deliver the surrender you had been hoping for. Your intelligence staff have broken a range of Japanese codes and ciphers, so you are able to build a pic-ture of events in Tokyo in the wake of the Soviet declaration of war.

To the peace faction within the Japanese leadership, this had seemed the moment to seize the initiative from the hardliners and force the acceptance of unconditional surrender, as President Truman had demanded at the Allies' Potsdam Conference a few weeks earlier,

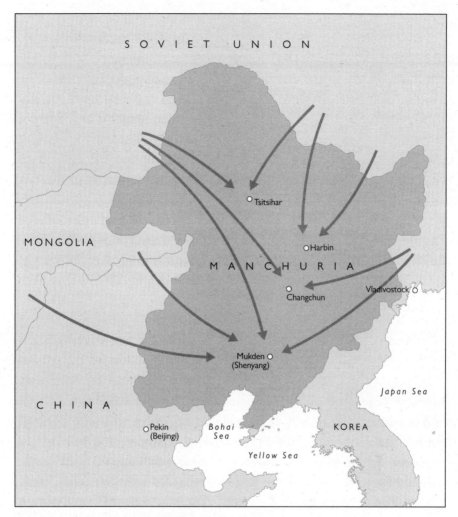

Map 2: Soviet Invasion of Manchuria

with the proviso that Emperor Hirohito's status and position be maintained (see Source 5).

Hirohito had broadly accepted these terms at the time of Potsdam, but the hardliners had raised objections. Following the Soviet attack, however, these positions began to shift a little, and at an emergency Supreme War Council meeting on 16 August, Prime Minister Kantarō Suzuki attempted once again to persuade the military chiefs to agree to the Potsdam offer. The War Minister, General Korechika Anami – regarded by many as the 'most powerful man in Japan' apart from the

Proclamation Defining Terms for Japanese Surrender, Issued, at Potsdam, 26 July 1945

1. We – the President of the United States, the President of the National Government of the Republic of China, and the Prime Minister of Great Britain, representing the hundreds of millions of our countrymen, have conferred and agree that Japan shall be given an opportunity to end this war.

13. We call upon the government of Japan to proclaim now the unconditional surrender of all Japanese armed forces, and to provide proper and adequate assurances of their good faith in such action. The alternative for Japan is prompt and utter destruction.

Source 5: The Potsdam Proclamation

Emperor – refused point-blank, insisting that the other terms Japan had been demanding had to be adhered to. There was still the hope among the hardliners that in addition to securing the role of the Emperor, Japan would be able to withdraw unhindered back to the homelands, avoid occupation and carry out its own war crime trials. (All of this is, in reality, totally unacceptable to the US government and public.) The meeting broke up in deadlock. It appears as though the Soviet attack did not shift the balance of power sufficiently and the war will go on.

Suzuki and the Foreign Minister, Shigenori Tōgō, then tried to persuade the Emperor himself to intervene directly, but Hirohito dithered until the following day: it was too late. Overnight the Prime Minister was assassinated by middle-ranking ultra-nationalist officers led by Major Kenji Hatanaka; Tōgō was also captured, while the other moderates fled. Anami and his hard-line colleagues resolved to double-down and fight on, and kept the hapless Hirohito at arm's length to avoid implicating him in any way. Although the position against the Soviet Union had all but collapsed by the end of August, and Stalin's troops had occupied South Sakhalin, the Kuril Islands and were advancing into northern Korea, the Japanese military chiefs resolved to carry on – they believe a final showdown on the Japanese homelands will simply be beyond the maritime capabilities of the USSR, while the decadent Americans will never have the stomach for a bloody war of attrition on the scale necessary to defeat Japan in her homelands. The military's *Ketsu-Go* plan to defend the Japanese home islands will therefore be put into operation, come what may. Some fanciful extremists even cling to the belief that just as Mongol invasion fleets had been destroyed by divine winds or *kamikazes* (typhoons) in the thirteenth century, so divine intervention might wipe out an American amphibious force in 1945.

When the news breaks that a *coup d'état* has in effect occurred in Tokyo, with the hardliners winning the day, you, Truman and Admiral Ernie King, head of the US Navy, are filled with dismay. It seems as though you will now have to commit to Olympic, even if you cling to the hope that sense might soon prevail in Tokyo, or that the blockade or air bombardment will force Japan to come to terms first; however, that now seems very unlikely to you. Preparations are ramped up across the western Pacific, and the US begins assembling a vast armada of ships, troops and aircraft to spearhead the invasion. Some 800,000 military personnel are allocated to Olympic alone, and in excess of one million more for the follow-up Operation Coronet.

For Olympic, the US Sixth Army, under the command of sixty-four-year-old General Walter Krueger, will provide the landing force of around fourteen divisions, grouped into three separate corps-level landings along the southern coast of Kyushu. The plan is for the landing forces to push sufficiently inland and occupy the southern half of Kyushu, which would then be used as a launching point for the main invasion of Honshu in March 1946 (Operation Coronet), to capture Tokyo itself.

Operation Olympic's invasion force will be escorted by the US Fifth Fleet under the command of Admiral Ray Spruance, and comprises forty-two aircraft carriers and twenty-four battleships, with many escorts, supporting vessels and the amphibious invasion force itself. Whilst you and Truman are still keen to keep non-US ground forces out of Japan, conversely the US Navy is happy enough for Commonwealth naval units to take part in the invasion. The Royal Navy's British Pacific Fleet (a combined Commonwealth force) will indeed provide around a quarter of the air element of the naval task force. Overhead tactical air support is to be provided by Fifth, Seventh and Thirteenth US Air Forces, while the long-range heavy bombers of the United States Strategic Air Forces in the Pacific – which combines American aircraft (increasingly the new B-29 Superfortresses) and a British strategic bombing fleet known as Tiger Force, including the famous 617 'Dambusters' Squadron – will offer heavy bombardment while maintaining the pressure on Japan's urban centres. The strategic air forces are under the command of the European-bombing war veteran, the American General Carl Spaatz, while overall executive command of the invasion has been allocated to the tough, egotistical and high-profile General Douglas MacArthur. The invasion is set for 1 November, or X-Day.

AFTERMATH

Despite all the hardware and meticulous preparations, Allied commanders expected heavy casualties in Olympic, although forecasts varied wildly – from under 100,000 to more than half a million. Estimates are based on loss rates in Europe, intelligence assessments of Japanese strength and the nature of the fighting in Iwo Jima and Okinawa. Both Admiral King and the Pacific Fleet's Commander-in-Chief, Admiral Chester Nimitz, are opposed to the invasion, gravely concerned about the possible losses at sea, while you too, though acknowledging Olympic's necessity, are fearful of the potential losses. Only MacArthur seems upbeat.

The Japanese were neither unaware nor unprepared for the invasion; they had accurately predicted what the Americans would do and where they would do it. Frantic activity takes place on Kyushu in the late summer of 1945, with fortifications under construction, underground bunkers for munitions and supplies established, and the careful husbanding of military equipment and ordnance. By the time of Olympic there are more than 850,000 military personnel on Kyushu, equipped with the best that Japan can provide at the time, although that is not much. Many units are seriously under-resourced and lack ammunition; many more have received scant training. Nonetheless, in Tokyo many in the government, even some hardliners, now realise that everything depends on Kyushu; a stout and costly campaign might compel the Americans to offer better terms. If it fails, there will be little left to offer any meaningful future resistance.

The Japanese Fifth Air Fleet in Kyushu has mustered many thousands of aircraft for *kamikaze* suicide attacks (over 30 per cent more than the Allied estimate of 7,000). In addition, a major propaganda campaign to mobilise the civilian population has been initiated, chillingly called 'The Glorious Death of One Hundred Million', which aims to encourage the unarmed people of Japan to sacrifice themselves in near-suicidal attacks on Allied forces. The Japanese navy, though now a shadow of its former self, can still offer midget submarines and suicide boats, alongside the seemingly most desperate measure: suicide divers equipped with mines.

For a fleeting moment in early October, Japanese hopes are raised when a typhoon does indeed hit American preparations with enough severity to delay the invasion by a number of weeks. Although this delay might ultimately have had some significant impact on Coronet by pushing back the start date to April 1946, when weather and the climate might well have hindered mobile operations in Honshu, in reality it merely delays the inevitable.

X-Day for Olympic is pushed back to 12 December and, after a massive air and naval bombardment, the invasion begins against an initially furious response from the desperate Japanese defenders. The *kamikaze* aircraft inflict some serious losses on Allied ships, but after two to three days the effort diminishes; Japanese aircraft prove increasingly unable to penetrate up to the amphibious landing ships, their primary targets, due to the intensity of Allied air cover, supported by radar-equipped B-17s flying early-warning missions high above. The Royal Navy's fully armoured aircraft carriers also prove far less susceptible to suicide attacks – US carriers had wooden flight decks, which increased the range and aircraft handling, but at the cost of protective armour. The Japanese navy's attempts to stall the invasion prove even less successful as the Allied anti-submarine forces, well versed in such operations in the Atlantic, largely hold off the threat. In the skies Allied air forces pummel Kyushu's defenders remorselessly, putting into practice many of the lessons learned in Europe and across the Pacific.

Japanese strategy is based on allowing the Americans control of the beaches and fighting them inland, away from Allied naval gunfire support. For a number of days the fighting is intense and losses mount up, although the major threat from Japanese civilians, who are dispirited, traumatised and increasingly war-weary, never materialises. The *kamikaze* threat also diminishes further as fuel runs low for the Japanese aircraft, and there is an increasing number of aborted missions as pilots start to lose faith. Japanese ground forces begin to fall back, and fewer show the desire to fight to the end than had been imagined by their bullish commanders in Tokyo. Nevertheless, it is a hard slog and Allied morale is also an issue, with many troops displaying sardonic and pessimistic tendencies; few relish the campaign.

The American forces eventually begin to push back the defenders at a quickening rate in late December, and Japanese commanders on the spot begin to worry about their ability to continue the fight much longer. Even in Tokyo concern is growing. Deaths from the B-29 raids are mounting alarmingly, now upwards of 750,000, and urban life in Japan has all but ceased to exist. Military production has collapsed and *Ketsu-Go* is showing no likelihood of extracting any concessions from the Americans. Matters take another dramatic downturn when, on 14 January 1946, news breaks that the Soviet Union has begun landing troops on Japan's principal northern island of Hokkaido.

Although a much more limited operation, and somewhat improvised compared to the grand scale of Olympic, the Soviet attack throws Tokyo into panic. Although there is little prospect of the Soviets managing anything on scale of the American invasion, Japan's diminishing strategic options across the board provoke further political convulsions in Tokyo.

On 20 January 1946 Admiral Mitsumasa Yonai, the Navy Minister, finally breaks ranks with the hardliners and openly pushes for surrender, much to the fury of General Anami. But when the Privy Council President, Kiichirō Hiranuma, also insists that the time had come to accept reality, the mood changes. The War Cabinet demands that Hirohito, isolated for many months, now be properly consulted. Realising that this is the end, Anami retreats to his office later that day and leaves a final note in which he states: 'I – with my death – humbly apologise to the Emperor for the great crime.' He then commits *seppuku.** Opposition to the peace party quickly melts away. With Hirohito's grateful consent, Hiranuma sends word to Washington that Japan wishes to surrender, unconditionally – save for securing the position of the Emperor, even if just as a constitutional monarch.

Three days later the surrender document is signed and fighting ceases, although Stalin's troops complete their occupation of Hokkaido. On 4 February the American battleship USS *Missouri* drops anchor in Tokyo Bay, and Douglas MacArthur accepts the formal surrender of Imperial Japan at a ceremony also attended by Vyacheslav Molotov, the Soviet Foreign Minister – a sign of things to come. The Second World War may have come to an end, but the new strategic environment in North-East Asia is stark indeed: as part of the settlement, Stalin retains both Hokkaido and the entire Korean peninsula, alongside South Sakhalin and the Kuril Islands. The Asia-Pacific theatre is soon to be gripped by the Cold War, just as Europe has been.

THE DECISION

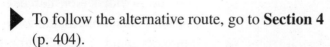 To follow the alternative route, go to **Section 4** (p. 404).

OR

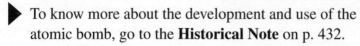 To know more about the development and use of the atomic bomb, go to the **Historical Note** on p. 432.

---

* Anami actually did this on 15 August 1945 after signing the surrender document.

SECTION 6
THE MANCHURIAN GAMBIT

2300 (Trans-Baikal time), 8 August 1945, the People's Commissariat for Foreign Affairs, Moscow

A difficult meeting is about to begin in the USSR's Foreign Ministry in Moscow. The Japanese ambassador in Moscow, Naotake Satō, has been summoned for urgent talks with you, *Vyacheslav Molotov*, the USSR's Foreign Minister. You have been Stalin's foreign-affairs man since 1939 and have been intimately involved with the major diplomatic events of the war, from the signing of the Russo-German Pact in 1939 through to the Potsdam Conference of just a few weeks ago.

8. Vyacheslav Molotov

At that conference Stalin had confirmed that the USSR would honour its previously agreed commitment to declare war on Japan three months after the ending of the war in Europe. That moment has now come, and you have summoned Sato to give him the bad news. Tokyo had been pinning its hopes on the USSR acting as a peace broker between Japan and the USA; you are about to extinguish that flicker of hope. You suspect that the Soviet declaration of war will force Tokyo's surrender, although not before

Stalin's troops have seized valuable territories and assets in North-East Asia. Just over an hour after you deliver the declaration of war to Sato, Soviet forces begin a three-pronged attack on Japanese-held territories in North-East Asia, predominantly Manchuria. The Second World War is surely entering its endgame.

The Soviet commander in the Far East, Marshal Aleksandr Vasilevsky, heads a vast force of 1.5 million troops equipped with 5,500 tanks, 27,000 artillery pieces and 3,700 aircraft. Supported by the experience of fighting the Axis forces in Europe and with finely honed operational doctrine and methods, the Soviets simply overwhelm the Kwantung Army. The Japanese are poorly equipped, with obsolete and inadequate tanks, artillery and aircraft, and many troops are raw recruits. Within days Soviet forces have smashed through Japanese defences and are soon pressing on relentlessly towards their objectives across Manchuria and Mongolia, even scooping up the Emperor of Manchukuo, who had previously acted as the last Emperor of China. In mid-August Soviet forces launch amphibious landings on the north Korean peninsula, the Kuril Islands and South Sakhalin.

In Washington, President Truman and his senior staff have been awaiting the impact of the invasion of Manchuria on Tokyo. Truman has held back the atomic bomb for a few weeks in the hope that the Soviet invasion, alongside the ever-tightening blockade and remorseless conventional bombing, would force Tokyo into surrender before he has to commit the new weapon; he remains quite willing to use the bomb, however, if Japan does not capitulate very quickly.

AFTERMATH

That the Soviet Union had declared war was not a complete surprise to Tokyo, or to General Otozō Yamada and his staff, commanding the Japanese Kwantung Army in Manchukuo, for they had been planning for just such an eventuality since the spring of 1945. But the Army General Staff in Tokyo had not held out much hope for the Kwantung Army, as many of Yamada's best troops had been redeployed to strengthen Japan's defences in the Pacific theatre. Convinced that the

USSR would not attack so quickly, Yamada's forces were also caught in the middle of a major redeployment, which had been initiated in a desperate effort to shore up the defences of the region. Very quickly the Kwantung Army collapsed.

In Tokyo the Japanese military leadership was in chaos and despair. The hardliners – those determined to fight on, despite the dire position they found themselves in – had based their obdurate resistance on the assumption that they could still offer some form of determined military opposition to any enemy incursion into the Japanese homelands. They cared little for the parlous situation of the civilian population, who had been relentlessly bombed since the spring of 1945, but they did pin a great deal of hope on offering fierce resistance to any American landings – what was known as the *Ketsu-Go* plan. Such a strategy was now exposed as hollow and meaningless, following the Soviet Union's attack on 8 August – there was simply no way that a worthwhile endgame strategy was possible, with so much military force arrayed against them, and when the feebleness of Japan's remaining armed forces was being so ruthlessly exposed.

Hard-line military opposition in Tokyo to any attempt to surrender on the Potsdam terms began to melt away in mid-August. One senior commander noted, 'upon hearing of the Soviet entry into the war, I felt that our chances were gone'. Admiral Toyoda, Chief of the Navy General Staff, likewise now conceded defeat. Following the Soviet attack, he later recorded, 'it became impossible for us to map any reasonable operation plan' – 'It was time for us to accept the terms of the Potsdam Declaration.' At an emergency meeting of the Supreme War Council on 18 August 1945, just as news of Soviet amphibious landings on the Korean peninsula broke, Prime Minister Kantarō Suzuki pressed the army chief General Korechika Anami and his hardliners to accede to the inevitable: that *Ketso-Go* was unworkable. There was also now the possibility that Japan might be invaded by the communist USSR, a fate worse even than occupation by the Americans. The army vacillated until Suzuki played his trump card, the unexpected appearance of Emperor Hirohito at the meeting. The Emperor tipped the balance in favour of surrendering by pressing for peace before it was too late.

A last-minute attempted *coup d'état* by middle-ranking army fanatics to prevent the act of surrender being followed through was defeated by the intervention of the loyal, fatalistic, if fanatical General Anami: indeed, shortly after stopping the rebellion, he committed suicide. Suzuki and Foreign Minister Tōgō contacted Washington and accepted the terms of Truman's Potsdam statement, although they requested clarity on the post-war position of the Emperor. In Washington, Truman was enthusiastic and Secretary of State James Byrnes – now concerned about further Soviet incursions into North-East Asia and the Pacific region – acquiesced to recognising the position of Hirohito as a constitutional monarch. On 22 August, Hirohito made a historic broadcast calling on the Japanese people to lay down their arms in the name of peace. It was an implicit, rather than explicit declaration of surrender, but nonetheless the Second World War came to an end at a formal ceremony on 9 September on board the American battleship USS *Missouri* in Tokyo Bay. In the Mariana Islands, hundreds of miles to the south, Colonel Paul Tibbets and his 509th Composite Group of the United States Army Air Force were stood down and their mysterious payloads disassembled, in readiness for transport back to the USA.

THE DECISION

▶ To follow the alternative route, go to **Section 7** (p. 426).

OR

▶ To know more about the development and use of the atomic bomb, go to the **Historical Note** on p. 432.

SECTION 7
HIROSHIMA AND NAGASAKI

0815, 6 August 1945, 31,000 feet above Hiroshima

The world is about to enter the age of atomic warfare. A Boeing B-29 Superfortress, *Enola Gay*, commanded by you, **Colonel Paul Tibbets** of the 509th Composite Group of the United States Army Air Force, is flying at a height of 31,000 feet. The aeroplane that you command is carrying the atomic bomb 'Little Boy', a uranium-235-based device, the first atomic weapon that is going to be used in anger in history. At thirty years old, you have been in the US Army's air forces since 1937 and have served in Europe, as well as being closely involved in the development of the B-29. Now you have been given this momentous mission.

President Truman himself, backed by his senior commanders, had given the go-ahead for the use of the atomic bombs some two weeks earlier, while the direct instruction (Operations Order No. 35) had been issued to your unit yesterday – the target is to be the Japanese city of Hiroshima.

9. Colonel Paul Tibbets

```
                        509TH COMPOSITE GROUP
                    Office of the Operations Officer
                      APO 247, c/o Postmaster
                      San Francisco, California

                                              5 August 1945

OPERATIONS ORDER )

NUMBER        35 )
                                    Out of Sacks:  Weather at 2230
Date of Mission:  6 August 1945                    Strike at 2330

Briefings:     : See below           Mess       :  2315 to 0115

Take-off:  Weather Ships at 0200 (Approx)   Lunches    :  39 at 2330
           Strike Ships at 0300 (Approx)                  52 at 0030

                                     Trucks     :  3 at 0015
                                                   4 at 0115
```

A/C NO.	VICTOR NO.	APCO	CREW SUBS	PASSENGERS (To Follow)
Weather Mission:				
298	83	Taylor		
303	71	Wilson		
301	85	Eatherly		
302	72	Alternate A/C		
Combat Strike:				
292	82	Tibbets	As Briefed	
353	89	Sweeney		
291	91	Marquardt		
354	90	McKnight		
304	88	Alternate for Marquardt		

```
GAS:  #82 - 7000 gals.
      All others - 7400 gals.

AMMUNITION:  1000 rds/gun in all A/C.

BOMBS:  Special.

CAMERAS:  K18 in #82 and #90.  Other installations per verbal orders.

RELIGIOUS SERVICES:  Catholic at 2200
                     Protestant at 2230

BRIEFINGS:

   Weather Ships
      General Briefing in Combat Crew Lounge at 2300.
      Special Briefings at 2330 as follows:
         AC and Pilots in Combat Crew Lounge
         Nav - Radar Operators in Library
         Radio Operators in Communications
         Flight Engineers in Operations
      Mess at 2330
      Trucks at 0015

   Strike Mission:
      General Briefing in Combat Crew Lounge at 2400
      Special Briefings at 0030 as follows:
         AC and Pilots in Combat Crew Lounge
         Nav and Radar Operators in Library
         Radio Operators at Communications
         Flight Engineers at Operations
      Mess at 0030
      Trucks at 0115    NOTE:  Lt McKnight's crew need not attend briefings.

                                         JAMES I. HOPKINS, JR,
                                         Major, Air Corps,
                                         Operations Officer.
```

10. Operations Order No. 35

11. The Enola Gay

You had arranged for the aircraft chosen for the mission to be named *Enola Gay* after your mother.

You know full well the implications of your mission, but your central thought is to get the job done and, hopefully, bring the war to an end.

You order the bomb's release, and the weapon descends for around forty seconds before detonating at a height of a little under 2,000 feet. Though it misses its intended target by 800 feet, the atomic bomb explodes with the force of sixteen kilotons of TNT, laying waste to an area of one-mile radius and causing severe fires across a radius of more than four miles. In just a few seconds some 80,000 people – around 30 per cent of the population of Hiroshima – are killed, while 70,000 more are injured. Over the next few days, weeks and months some 60,000 more people will perish from burns, wounds and radiation sickness.

AFTERMATH

The following day, 7 August, senior Japanese figures began conferring on the implications of what might have happened at Hiroshima – it was still unclear that it had been an atomic bomb. The hardliners, pre-dominantly the arch-militarists in the army, still rejected the idea of accepting peace on the terms of Truman's Potsdam Proclamation, which had called for Japan's unconditional surrender – there was, in any case, still some confusion as to whether the Americans would allow Emperor Hirohito to remain in place, even if not in political power. Even on 7 August some hardliners in Tokyo believed that a peace settlement founded on the maintenance of the Emperor as a divine ruler, Japanese self-withdrawal to the home islands, no enemy occupation of Japan, and war-crimes trials to be conducted by Japan itself, was still the basis for a cessation of hostilities. There was the continuing hope that the as-yet-neutral Soviet Union would act as an intermediary in these negotiations.

But the Emperor himself and the peace faction in the Japanese government had been shaken by the news of the Hiroshima bombing and wanted to push for a settlement. 'Now that things have come to this impasse, we must bow to the inevitable . . . we should lose no time in ending the war so [as] not to have another tragedy like this,' Hirohito supposedly told an advisor. Indeed, the intervention of the Emperor, however initially tentative, and probably driven by a desire to preserve the monarchy as much as anything else, sent a shockwave through his government. Nonetheless the Army stalled, by asking that they should all await the results of the investigation into what happened at Hiroshima, and the outcome of further efforts to get the Soviet Union to act as intermediaries.

When it became clearer that it had been an atomic bomb at Hiroshima, Prime Minister Kantarō Suzuki pressed for Tokyo to open negotiations with Washington as soon as possible, although he may still have harboured hopes for something better than the Americans appeared to be offering at Potsdam. Japan's resistance seemed to end, however, when in the final hours of 8 August – far from acting as brokers between Tokyo and Washington – the Soviet Union declared war on Japan and began the invasion of Japanese-occupied Manchuria.

The *coup de grâce* for Japan's lingering hopes came while this news was being debated by the Supreme War Council on the morning of 9 August; reports arrived of a second atomic bombing, this time at

Nagasaki. 'Fat Man', a plutonium bomb, was detonated a little after 1100. Although the bomb was technically more powerful than the Hiroshima device, it missed its intended target by some distance, resulting in the terrain absorbing some of the blast and destructive power. Nevertheless, somewhere between 35,000 and 40,000 people were killed instantly, with 60,000 further casualties. Tens of thousands more would die in the aftermath as a result of the bomb.

Yet even now the Supreme Council, and later a full Cabinet meeting, was split roughly evenly on whether to accept Truman's Potsdam Proclamation (with the proviso that the Emperor remain as head of state) or insist on the further points, as previously insisted upon by the hardliners. Truman, however, kept up the pressure by stating in a new announcement that the US would continue to use atomic bombs until Japan surrendered. To a degree he was bluffing, as the USA could at best muster one more bomb for use in mid-August, with three more ready in September and three further devices in October, all of them plutonium bombs. However, Truman, having now seen photographs of the effects of the atomic bombs, was hesitating about their ongoing use without his express and very specific agreement (see Source 6).

According to one source: '[Truman] said he had given orders to stop atomic bombing. He said the thought of wiping out another 100,000 people was too horrible. He didn't like the idea of killing he said, "all those kids".'

In the late evening of 9 August Hirohito was again called upon to intervene and gave his view that surrender – on the single proviso that the monarchy be maintained – was now the only course to take: 'I have given serious thought to the situation prevailing at home and abroad and have concluded that continuing the war can only mean destruction for the nation and prolongation of bloodshed and cruelty . . . I swallow my tears and give my sanction to the proposal to accept the Allied proclamation . . .' One brave attendee asked whether the Emperor himself would also accept some responsibility for the defeat. Once Hirohito had departed, the Cabinet finally accepted defeat and asked the Allies about a formal surrender.

MEMORANDUM TO: Chief of Staff. The next bomb of the implosion type had been scheduled to be ready for delivery on the target on the first good weather after 24 August 1945. We have gained 4 days in manufacture and expect to ship from New Mexico on 12 or 13 August the final components. Providing there are no unforeseen difficulties in manufacture, in transportation to the theatre or after arrival in the theatre, the bomb should be ready for delivery on the first suitable weather after 17 or 18 August.

L. R. Groves, Major General, USA.

8/10/45
It is not to be released over Japan without express authority from the President
[signed] G C Marshall

Source 6: Memo from Groves to Marshall, 10 August 1945

In Washington the news was received with great relief, although the hard-line Secretary of State James Byrnes then demanded that the Japanese be very clear that they had accepted that the Emperor would in future only act as a constitutional monarch. This was finally agreed by the Japanese Cabinet on 14 August, which then prompted an attempted *coup d'état* that was, however, soon quashed.

On 15 August 1945 Hirohito spoke to the Japanese people – the first time that the vast majority of them had ever heard his voice. They were shocked to learn that the war was over and that Japan had in effect been defeated, although the Emperor carefully avoided use of the word 'surrender'. On 2 September 1945 a formal surrender ceremony took place on board the USS *Missouri* in Tokyo Bay, witnessed by General Douglas MacArthur for the USA. The Second World War was over.

THE DECISION

▶ To follow the alternative route, go to **Section 6** (p. 422).

OR

▶ To know more about the development and use of the atomic bomb, see the **Historical Note** below.

HISTORICAL NOTE

The road to the development and use of the atomic bomb dated back to before the outbreak of the Second World War, when the Nazis' warped racial policies set in motion the events that would hand the initiative to the Allies. The historical path in this chapter was for the Bohrs to reject any attempt to engage with Heisenberg, even if in the hope of deflecting their old friend from the path that he was embarked upon (from Section 1, go to Section 2). By October 1941 Vannevar Bush, prompted by a variety of factors and influences, had identified the necessity of urgently investing major resources in what became the Manhattan Project (from Section 2, go to Section 4). It is doubtful that Bohr's news of his meeting with Heisenberg was an immediate driver in Bush's decision to ask Roosevelt and Wallace for the necessary backing for fission research, but intelligence and evidence were mounting that the USA had to act.

Once the decision to develop the bomb had been made, the last issue was whether to use it, and when. It remains one of the most contentious and debated topics of the closing stages of the war – was it necessary or wise to use the atomic bomb, particularly in the circumstances of the Asia-Pacific War in July / August 1945? Since the 1960s some historians have argued that the bomb's use was politically motivated to intimidate the USSR – if the USA had the bomb and the Soviet Union did not, Stalin would surely have to toe the line. This could only be achieved by demonstrating the bomb in action, and that was why it was used against Japan when that country was trying to surrender.

Yet the terms on which Japan was seeking to end the war in July 1945 were simply unacceptable to Washington. Truman took the decision to use the bomb to try and end the war as quickly as possible, although it is true that some in Washington also saw diplomatic benefits from its use (from Section 4, go to Section 7). Whether it was the correct strategic option, or indeed moral choice, is still open to question. It is still unclear whether the bomb, or the Soviet declaration of war, or both, were the reasons that prompted Tokyo to end the war in August 1945.

What if Bohr had debated the atomic-weapon issue with Heisenberg in 1941? Was a German bomb a possibility? By driving out many of the theoretical physicists that would be needed to develop an atomic bomb, the Nazis had in fact left Heisenberg with too little support when it came to the Third Reich's atomic power and weapons programme. Even if Niels Bohr had somehow alerted Heisenberg to some of the theoretical problems that he had not yet engaged with (from Section 1, go to Section 3), it is difficult to see how the Germans could have overcome other problems, such as access to uranium and the diversion of resources into Wernher von Braun's rocket projects. Germany's atomic programme at the end of the war was heading towards atomic power, but weapons were a distant prospect. There is also little doubt that if the war had gone on into the summer and autumn of 1945, the USA would have deployed the atomic bomb against Hitler's Germany.

What would the consequences have been, if Bush had not convinced Roosevelt about 'fission research' in October 1941? It was in truth quite an ask, because many were still sceptical about the atomic-bomb idea, despite British insistence to the contrary. There were also other, clearly more realistic weapons programmes that the USA could allocate resources to in the autumn of 1941, and it has of course to be remembered that the Americans were at peace when Roosevelt agreed to Bush's request to press forward with the development of the atomic bomb. Even then it was a close-run thing that the atomic bomb would be ready before the war ended, so any delay in funding in the autumn of 1941 could well have narrowed Truman's options for forcing Japan's surrender in July 1945 by depriving him of the bomb (from

Section 2, go to Section 5). If the Soviet declaration of war, and the US bombing and blockade campaigns, had not pushed Tokyo over the edge, then Downfall (Olympic / Coronet) might well have been inevitable.

If Truman had hesitated over using the bomb in July 1945, the Soviet Union's declaration of war on Japan on 8 August would have been the best hope of forcing Tokyo to surrender (from Section 4, go to Section 6). Indeed, some historians have argued that this was more important than the bomb in compelling Tokyo's capitulation. Without the bomb, would the Soviet attack have been enough? It is possible, for the hardliners in Tokyo were mainly from the army and their credibility and hopes might have been shattered by the rapid collapse in Manchuria, but some senior military leaders had already written off the Japanese army there, so it is by no means certain that it would have had such an effect.

When Colonel Paul Tibbets in the *Enola Gay* gave the order to drop the first atomic bomb, the world entered a new age of warfare. Though similar numbers of people had been killed in the firebombing of Tokyo in March 1945, the effects of a single device causing such devastation and destruction pushed warfare into a new, appallingly dangerous era.

Margrethe Bohr was married to Niels for fifty years and aided him in his writing and research for much of that period. She died at the age of ninety-six in Copenhagen in 1984.

Vannevar Bush continued his work as director of the Office of Scientific Research and Development until the war's end and hoped to develop an equivalent peacetime body. He was somewhat frustrated in this, and took up a variety of posts at leading American companies and institutions in the post-war world.

Henry Stimson – tough, resolute and uncompromising – played a vital role in American foreign and defence policy for much of the first half of the twentieth century, serving in both Republican and Democrat administrations. He retired in 1945 at the age of seventy-eight.

General George Marshall played a vital role throughout the war acting as Roosevelt's key military advisor. He passed up the opportunity to take a major field command, though deep down he probably wanted the Overlord command. Highly regarded for his determined grip and clear sighted vision he was an architect of Allied victory.

Vyacheslav Molotov remains a key figure in the history of the USSR, although he fell from grace after the Second World War. He remained loyal to Stalin's legacy and opposed the Khrushchev policy of denouncing Stalinism in the 1950s.

Paul Tibbets is forever remembered as the pilot of the B-29 Superfortress *Enola Gay*. He remained in the US air forces until 1966 and never shirked from the responsibility, necessity and impact of his mission on 6 August 1945.

ACKNOWLEDGEMENTS

Special thanks are due to Howard Fuller, Spencer Jones, Andy Southall and Martyn Shipley for all the games and speculation over many years.

REFERENCES

Though there are many examples of alternative history books and novels, this is a selection of those I have found most interesting whilst writing this book.

Robert Cowley (ed.), *What If? Military Historians Imagine What Might Have Been,* (1999)

Robert Cowley (ed.), *More What If? Eminent Historians Imagine What Might Have Been,* (2001)

Niall Ferguson (ed.), *Virtual History: Alternatives and Counterfactuals,* (1997)

Andrew Roberts (ed.), *What Might Have Been: Leading Historians on Twelve 'What Ifs' of History,* (2004)

For more on the specific chapters:

Britain's Darkest Hour

John Lukacs, *Five Days in London May 1940,* (1999)

Nicholas Shakespeare, *Six Minutes in May: How Churchill Unexpectedly Became Prime Minister,* (2017)

The Mediterranean Gamble

Niall Barr, *Pendulum of War: Three Battles of El Alamein,* (2004)

John Baynes, *The Forgotten Victor: General Sir Richard O'Connor,* (1989)

Desmond Dinan, *The Politics of Persuasion: British Policy and French African Neutrality, 1940-42,* (1988)

Stalin's War

Geoffrey Roberts, *Stalin's War: From World War to Cold War, 1939-1953,* (2006)

Vojtech Mastny, 'Stalin and the Prospects of a Separate Peace in World War II', *The American Historical Review,* (December 1972)

Midway – Decision in the Pacific

Jonathan Parshall, *Shattered Sword: The Untold Story of the Battle of Midway,* (2005)

Stephen L Moore, *Pacific Payback: The Carrier Aviators Who Avenged Pearl Harbor at the Battle of Midway,* (2014)

Bomber Offensive

Mark Connelly, *Reaching for the Stars: A History of Bomber Command,* (2001)

Peter Gray, *The Leadership, Direction and Legitimacy of the RAF Bomber Offensive From Inception to 1945,* (2012)

Casablanca

Niall Barr, *Yanks and Limeys: Alliance Warfare in the Second World War,* (2015)

Walter Scott Dunn, *Second Front Now – 1943,* (1980)

Race to the Rhine

John Buckley, *Monty's Men: The British Army and the Liberation of Europe 1944-5,* (2013)

Sebastian Ritchie, *Arnhem: Myth and Reality – Airborne Warfare, Air Power and the Failure of Operation Market Garden,* (2011)

The Bomb

Jim Baggott, *The First War of Physics: The Secret History of the Atomic Bomb,* (2010).

J Samuel Walker, *Prompt and Utter Destruction: Truman and the Use of the Atomic Bombs Against Japan,* (2017 edition)

PERMISSIONS

1. BRITAIN'S DARKEST HOUR

Images

1 © Popperfoto/Getty
2 https://commons.wikimedia. org/wiki/File:0929_fc- churchill-halifax.jpg, public domain
3 © Hulton Deutsch/Getty
4 © Popperfoto/Getty
5 © Central Press/Stringer/ Getty
6 George Grantham Bain collection, public domain
7 Dutch National Archives, public domain
8 © ANL/Shutterstock
9 © Keystone/Stringer/Getty
10 © Corbis Historical/Getty
11 Crown copyright expired

2. THE MEDITERRANEAN GAMBLE

Images

1 Crown copyright expired
2 © Fotosearch/Stringer/Getty

3 Public domain
4 National Archives of the Netherlands, public domain
5 https://tr.wikiquote.org/wiki/ Dosya:Ismet_Inonu_portrait. jpg, public domain
6 Public domain
7 National Archives of the Netherlands, public domain
8 © IWM H 39144
9 © Express/Stringer/Getty
10 https://commons.wikimedia. org/wiki/File:- Brigadier_W_H_E_Gott.jpg, crown copyright expired

Maps

1 Redrawn based on original © jackaranga 2007
6 Redrawn based on original © Memnon335bc 2011

3. STALIN'S WAR

Images

1 © Popperfoto/Getty
2 https://en.wikipedia.org/wiki/ Nikolai_Vlasik#/media/

File:Н.С._Власик_в_
рабочем_кабинете._
Конец_1930-х_годов.gif,
public domain

3 Public domain
4 Public domain
5 Public domain
6 © Laski Diffusion/Getty
7 Public domain
8 Public domain
9 Public domain
10 New York Times, public
domain
11 Bertil Norberg/Pressens Bild,
https://commons.wikimedia.
org/wiki/File:Press_
photo_of_Christian_
G%C3%BCnther_1940.jpg,
public domain

Maps

1 Redrawn based on original ©
Gdr 2005
7 Redrawn based on original ©
Gdr 2005
8 Redrawn based on original ©
Gdr 2005

4. MIDWAY

Images

1 US Naval Institute, https://
en.wikipedia.org/wiki/
Lynde_D._McCormick#/
media/File:ADM_
McCormick,_Lynde_D.jpg,
public domain
2 Public domain

3 National Museum of the US
Navy https://commons.
wikimedia.org/wiki/
File:80-G-317645_
(18520533280).jpg, public
domain
4 Naval History and Heritage
Command, public domain
5 Naval History and Heritage
Command, public domain
6 Naval History and Heritage
Command, public domain
7 Naval History and Heritage
Command, public domain
8 Public domain

5. BOMBER OFFENSIVE

Images

1 © FPG/ Getty
2 Tangmere Military Aviation
Museum, public domain
3 © IWM CH 9504
4 Avro Heritage Museum,
public domain
5 © Popperfoto/Getty
6 Public domain
7 © Leonard McCombe/
Stringer/Getty
8 Canadian Warplane Heritage
Museum, public domain
9 Crown copyright expired
10 Crown copyright expired
11 German photo library/
Richard Peter Sr., public
domain
12 © Popperfoto/Getty

6. CASABLANCA

Images

1 Public domain
2 © Galerie Bilderwelt/Getty
3 National Archives of the Netherlands, public domain
4 Public domain
5 German Federal Archive, public domain
6 © Historical/Getty
7 © Hulton Archive/Stringer/Getty
8 Department of the Army Harry S. Truman Library & Museum, copyright expired

7. RACE TO THE RHINE

Images

1 Crown copyright expired
2 ibiblio.org, public domain
3 Crown copyright expired
4 © IWM/Getty
5 https://en.wikipedia.org/wiki/Edgar_Williams#/media/File:EdgarWilliams.jpg, public domain
6 U.S. Army file photo, http://www.usarpac.army.mil/history/cgbios/images/33-parks_l.jpg, public domain

Maps

1 Crown copyright expired
2 © warfaremagazine.co.uk, taken from *Major and Mrs Holt, Major And Mrs Holt's Battlefield Guide: Operation Market Garden*, 3rd edition (2013)

8. THE BOMB

Images

1 Niels Bohr Institute, public domain
2 Fermilab, U.S. Department of Energy, public domain
3 © Keystone-France/Gamma-Keystone/Getty Images
4 Bassano Ltd/National Portrait Gallery, Australia, expired crown copyright
5 © Ullstein Bild Dtl./Getty
6 © Bettmann/Getty
7 Image courtesy of the US Army's Center of Military History, public domain
8 © Paul Popper/Popperfoto/Getty
9 © Bettmann/Getty
10 Photo taken by U.S. Government military personnel on Tinian Island in 1945, public domain
11 © Keystone/Stringer/Getty